Australian Perspectives on Global Air and Space Power
Past, Present, Future

Edited by Nicole Townsend, Kus Pandey and Jarrod Pendlebury

Routledge
Taylor & Francis Group

LONDON AND NEW YORK

First published 2023
by Routledge
4 Park Square, Milton Park, Abingdon, Oxon OX14 4RN

and by Routledge
605 Third Avenue, New York, NY 10158

Routledge is an imprint of the Taylor & Francis Group, an informa business

British Library Cataloguing-in-Publication Data
A catalogue record for this book is available from the British Library

ISBN: 978-1-032-13740-7 (hbk)
ISBN: 978-1-032-13739-1 (pbk)
ISBN: 978-1-003-23065-6 (ebk)

DOI: 10.4324/9781003230656

Typeset in Times New Roman
by Apex CoVantage, LLC

Australian Perspectives on Global Air and Space Power

This book surveys historical and emerging global air and space power issues and provides a multidisciplinary understanding of the application of air and space power in the past and present, while exploring potential future challenges that global air forces may face.

Bringing together leading and emerging academics, professionals, and military personnel from Australia within the field of air and space power, this edited collection traces the evolution of technological innovations, as well as the ethical and cultural frameworks which have informed the development of air and space power in the 20th and 21st centuries, and contemplates the future. It covers topics such as insurgents' use of drones, the ethics of air strikes, the privatisation of air power, the historical trajectory of air power strategy, and the sociological implications of an 'air force' identity. While many of the chapters use Australian-based case studies for their analysis, they have broader applicability to a global readership, and several chapters examine other nations' experiences, including those of the United States and the United Kingdom.

This accessible, illuminating book is an important addition to contemporary air and space power literature, and will be of great interest to students and scholars of air and space power, air warfare, military and international history, defence studies, and contemporary strategic studies, as well as military professionals.

Nicole Townsend is currently completing a PhD in History at the University of New South Wales, Canberra. Her thesis focuses on Australia's war in the Mediterranean during the Second World War, and she has both presented and published more broadly on the Second World War. She currently works as a researcher on the Official History of Australian Operations in Iraq and Afghanistan, based at the Australian War Memorial.

Kus Pandey was the manager of the Australian Centre for the Study of Armed Conflict and Society (ACSACS) at the Australian Defence Force Academy campus of the University of New South Wales from 2018 to 2019. She was awarded an Australia Day Medallion by the Department of Defence for her contribution to ACSACS, with specific recognition of her support of the Sir James Rowland Seminars.

Jarrod Pendlebury is an honorary research fellow with The King's School Institute and a senior strategist in the Royal Australian Air Force. His research interests focus on the intersection of sociology and strategy, with particular emphasis on air power, and he has been published widely in Australia, Europe, and the United States.

Contents

Contributor Biographies

Kristen Alexander is Adjunct Associate Lecturer at UNSW Canberra. She specialises in Australian aviation history and holds a PhD in History from UNSW Canberra. In 2022, Dr Alexander was awarded the Australian War Memorial's Bryan Gandevia Prize for Australian military-medical history for her PhD thesis, 'Emotions of Captivity: Australian Airmen Prisoners of Stalag Luft III and their Families'. She has been published in Australia, the United Kingdom, and Japan, and two of her books, *Jack Davenport Beaufighter Leader* and *Australia's Few and the Battle of Britain*, were included on the Royal Australian Air Force's Chief of Air Force reading list in 2010 and 2015, respectively. Currently, Dr Alexander is writing a book based on her PhD thesis and revising her first biography*, Clive Caldwell Air Ace*.

Deane-Peter Baker is Associate Professor of International and Political Studies in the School of Humanities and Social Science at UNSW Canberra. He serves as Executive Director of Security & Defence PLuS, a collaboration within the PLuS Alliance (Arizona State University, Kings College London, and UNSW). Baker is also a senior visiting research fellow in the Kings College London Centre for Military Ethics. He served as a panellist on the International Panel on the Regulation of Autonomous Weapons (IPRAW). His current area of focus is ethics and special operations, and he is a regular consultant to Australia's Special Operations Command. Recent publications include: *Should We Ban Killer Robots?* (Polity Press, 2022), *Morality and Ethics at War: Bridging the Gaps Between the Soldier and the State* (Bloomsbury Academic, 2020), and *Citizen Killings: Liberalism, State Policy and Moral Risk* (Bloomsbury Academic, 2016).

Jason Begley, CSM, joined the Royal Australian Air Force as a navigator in 1991, graduating from the Australian Defence Force Academy with a Bachelor of Arts with Honours in Politics in 1994 and the School of Air Navigation in 1996. Following conversion to the P-3C Orion in late 1996, he was posted to 10 Squadron; his subsequent career involved multiple operational flying tours on Orions and staff postings in Canberra supporting Air Force projects. Highlights include roles as the Australian Liaison to US Central Command's Combined

Theatre Electronic Warfare Coordination Cell at Al Udeid Air Base, Commanding Officer No. 10 Squadron, Director of Joint Effects at Headquarters Joint Operations Command, and Director of the Air and Space Power Centre. He is a graduate of the Australian Command and Staff College and National Security Fellowship at the Harvard John F. Kennedy School of Government. Currently, he is Director General Joint Command, Control, Communications and Computers in Joint Capabilities Group.

Jo Brick is a legal officer in the Royal Australian Air Force. She has served on several operational and staff appointments from the tactical to the strategic levels of the Australian Defence Force, including Staff Officer (Legal) to the Chief of the Defence Force and Legal Adviser to the former Chief of Air Force. Group Captain Brick is a graduate of the Australian Command and Staff Course – Class of 2014. She holds a Master of International Security Studies (Deakin University), a Master of Laws (Australian National University), and a Master of Military and Defence Studies (Advanced) (Hons) (Australian National University). She is a member of the Military Writers Guild, an associate editor for The Strategy Bridge, and an editor for The Central Blue. She was recently appointed as a non-resident fellow of the United States Marine Corps Krulak Center.

Tom Frame, AM, joined the RAN College, HMAS Creswell, as a 16-year-old junior entry cadet midshipman in January 1979. He served at sea and ashore, including a posting as Research Officer to the Chief of Naval Staff, and completed a PhD at UNSW Canberra in 1991 on the 1964 HMAS Voyager disaster. He resigned from the RAN to train for the Anglican ministry in 1993. After parish work in Australia and England, he was Bishop to the Australian Defence Force (2001–07) and then Director of St Mark's National Theological Centre (2007–14). He has been a visiting fellow in the School of Astronomy and Astrophysics at the Australian National University (2000–03); a Patron of the Armed Forces Federation of Australia (2002–06); a member of the Council of the Australian War Memorial (2004–07); and judge for the inaugural Prime Minister's Prize for Australian History (2007). He was appointed Director of the Australian Centre for the Study of Armed Conflict and Society at UNSW Canberra in July 2014. In July 2017, he became the inaugural Director of the Public Leadership Research Group—Howard Library. In July 2020, he was commissioned by the Chief of the Defence Force to produce a major study of allegations of misconduct by Australian Special Forces personnel in Afghanistan, 2005–13. His book *Veiled Valour: Australian Special Forces in Afghanistan and war crimes allegations*, was published by UNSW Press in 2022.

James-Andre Galam completed his Bachelor of Aeronautical Engineering (Honours) at the Australian Defence Force Academy in 2021. Following this, he was posted to Air Combat and Electronic Attack Systems Project Office in Royal Australian Air Force Base Amberley. He has a keen interest in space power and hopes to work with ADF space technology in the future.

Travis Hallen is a serving aviator in the Australian Defence Force. He joined the Royal Australian Air Force in 2000 as a direct entry navigator trainee after completing a Bachelor of Arts (Honours) in Japanese from the University of Queensland. After graduating Basic Navigator Course in 2002, Hallen started his professional career on the P-3C Orion Maritime Patrol Aircraft. Since then, he has completed a range of operational, flying, and staff roles. Wing Commander Hallen is a graduate of the US Air Force School of Advanced Air and Space Studies. He is a passionate advocate of professional military education and was a founding editor of the Williams Foundation blog 'The Central Blue'. In 2019, Wing Commander Hallen was appointed to the position of Air Staff Officer Plans and Operations at The Australian Embassy in Washington, D.C. He is currently an air power strategist on exchange with the US Air Force's Futures and Concepts Division.

Matt Hegarty, CSC, joined the Australian Defence Force Academy in 1988, graduating with a science degree. He graduated from No. 2 Flying Training School Royal Australian Air Force Base Pearce in 1992. He has served with 37 Squadron, Headquarters Logistics Command, 36 Squadron, Headquarters Air Lift Group, 86 Wing, 37 Squadron, Air Force Personnel Directorate, KC-30A Transition Team, and was Chief of Staff to the Chief of the Defence Force before becoming the Director Strategic Design in the office of Director General Strategy and Planning in Air Force Headquarters. In 2017, Air Commodore Hegarty was posted as the Commandant of the Australian Command and Staff College. He has completed several operational deployments, including as deputy Commander Joint Task Force 633. He is currently Director General Military Strategy in Strategic Policy Division. Air Commodore Hegarty's academic achievements include the completion of the Royal Australian Air Force's Aeronautical Integrated Systems Course (RAAF School of Air Navigation), a Master of Management in Defence Studies (University of Canberra) and a Master of Arts in Strategic Studies (Deakin University).

Amy Hestermann-Crane is an analyst in the Royal Australian Air Force. In 2020, Hestermann-Crane became the first enlisted Air Force space analyst within the Australian Space Operations Centre. She is a member of the Australian Capital Territory's Ministerial Advisory Council for Veterans and Their Families and the Royal Australian Air Force Women's Integrated Networking Group. Through these roles, she focuses on influencing policies to improve veteran and family life within the ACT, promote and provide educational opportunities, mentoring, and encourage engagement from Air Force and the broader community. Sergeant Hestermann-Crane has a passion for ethics and STEM, putting both interests to use as a member of the International Space Ethics Collaborative Research Team and a visiting instructor at the Australian Youth Space Academy. She has completed a Bachelor of Communication and Bachelor of Historical Inquiry and Practice and is working towards completing a Bachelor of Arts (Honours) and a Master in Space Operations.

Peter Hobbins is a historian of science, technology, and medicine. Head of Knowledge at the Australian National Maritime Museum, he is also an honorary associate in the Department of History at the University of Sydney and a visiting scholar at the State Library of New South Wales. Peter's research has spanned two centuries of Australian history, including snakebite remedies, scientific medicine, epidemics, and quarantine. He has also published extensively on defence technologies and military medicine, including troop transports, war neurosis, radar, and aviation medicine. Peter is currently writing his third book, which focuses on the history of aviation accidents in Australia.

Peter Hunter has over 25 years' experience in Australia's national security community, having worked across a broad range of positions in foreign and strategic policy, including the Department of Foreign Affairs and Trade, the Department of Prime Minister and Cabinet, and the Office of National Assessments. He has also served as adviser to the Minister for Defence and has undertaken diplomatic postings to Papua New Guinea and Kenya. Hunter commenced his career as an officer in the Royal Australian Air Force, where he continues to serve as a Wing Commander in the Air Force reserve. He recently completed a PhD on Australian air power strategy through the University of NSW.

Keirin Joyce, CSC, is the RPAS/UAS Chief Engineer for the Royal Australian Air Force and has been extensively involved with UAS development in the Australian Defence Force for the past 16 years. He was awarded the Conspicuous Service Cross in 2012 for enabling the integration of the uncrewed aerial systems capability in support of operations in Afghanistan.

Peter Layton is a visiting fellow at the Griffith Asia Institute, Griffith University; an associate fellow at the Royal United Services Institute; and a fellow of the Australian Security Leaders Climate Group. He has extensive aviation and defence experience and, for his work at the Pentagon on force structure matters, was awarded the US Secretary of Defense's Exceptional Public Service Medal. He has a doctorate from the University of New South Wales on grand strategy and has taught on the topic at the Eisenhower College, US National Defense University. For his academic studies, he was awarded a fellowship to the European University Institute, Fiesole, Italy. His research interests include grand strategy, national security policies particularly relating to middle powers, defence force structure concepts, and the impacts of emerging technology. He regularly contributes to the public policy debate on defence and foreign affairs issues and is the author of the book *Grand Strategy*. His articles, posts, and papers can be accessed at <https://peterlayton.academia. edu/research#papers>.

Michael Molkentin is the coordinator of English, drama, and languages at Shellharbour Anglican College and an Adjunct Lecturer in History in the School of Humanities and Social Sciences at UNSW Canberra. He has a PhD in History from UNSW and is the author of four books on aviation history. His critically acclaimed latest work *Anzac and Aviator* (Allen & Unwin, 2019) is a biographical history of the Australian airman Sir Ross Smith.

Kus Pandey is the former manager of the Australian Centre for the Study of Armed Conflict and Society, based at the University of New South Wales, Canberra/the Australian Defence Force Academy. For her work with and dedication to the Centre, she was awarded an Australia Day Medallion in 2019 as part of the Australia Day Honours. She currently works at the Australian National University and has previously managed a project on moral injury sustained in combat.

Jarrod Pendlebury is a senior Air Force strategist who has held a wide range of appointments ranging from the tactical application of air power to the development of force structure options for the Australian Defence Force. As the Royal Australian Air Force's Director of Strategic Design, he led a small team responsible for the development of *The Air Force Strategy 2020* (AFSTRAT), which represented a milestone in the broadening of the Air Force's articulation of strategic value to the government beyond a narrow focus on the application of kinetic effects. He has wide experience as a pilot on the C-130H Hercules, C-17A Globemaster, and C-27J Spartan, and he commanded Number 35 Squadron Royal Australian Air Force during the introduction to type of the latter. As a pilot, he has vast operational experience in warlike and peacetime air operations, including in Iraq and Afghanistan and following the 2004 Boxing Day Tsunami. He commanded Australia's air response in support of the Philippine Government following Super Typhoon Haiyan/Yolanda in 2013. He holds a Bachelor of Arts in Politics and History, a Master of Human Rights, a Master of Philosophy in Military Strategy, and a PhD in Military Sociology. He is currently posted to New York City in a diplomatic role as Australia's Military Advisor to the United Nations.

Michael Spencer penned this chapter as an Officer Aviation (Maritime Patrol & Response) at the Air Power Development Centre, analysing the potential risks and opportunities posed by technology change drivers and disruptions to the future employment of air and space power. At the end of this posting, with a service career spanning 40 years, he transitioned to the Air Force Reserve to join the Defence COVID-19 Task Force and Air Force Headquarters projects for Remotely Piloted Aircraft Systems. His Air Force career has provided operational experiences in long-range maritime patrol, aircrew training, and weaponeering, and management experiences in international relations, project management in air and space systems acquisitions, space concepts development, and joint force capability integration. He is a project manager certified by the Australian Institute of Project Management, an associate fellow of the American Institute of Aeronautics & Astronautics, and a member of the Australia New Zealand Space Law Council.

Nicole Townsend is currently completing a PhD at the University of New South Wales, Canberra. Her thesis focuses on Australia's war in the Mediterranean during the Second World War. A Director of the Second World War Research Group, Asia Pacific, she specialises broadly in the Second World War. Currently, she works as a researcher on the Official History of Australian Operations in Iraq and Afghanistan, based at the Australian War Memorial.

Charles Vandepeer is a senior lecturer in intelligence and security studies at Charles Sturt University, Australia. His career has included service in the Royal Australian Air Force as an intelligence officer and with the Defence Science and Technology Organisation as a civilian defence operations research scientist. As a Squadron Leader, Charles served as the senior Air Force lead for Australian and Coalition analytic teams on deployment, gaining operational experience in the Middle East. Charles completed his PhD at the University of Adelaide, examining intelligence analysis and threat assessment within Australia, the United States, and the United Kingdom. Studying and completing his entire undergraduate degree at the University of the South Pacific (Fiji), Charles gained first-hand experience in understanding the power and influence that cultures, mindsets, and perspectives have on how we interpret and interact with the world around us. He is the author of the book *Applied Thinking for Intelligence Analysis* and a general readership book *Asking Good Questions*, and also runs the *Mind of War Project*, which examines the enduring human aspect of warfare (www.mindofwarproject.com/).

Christopher Wooding is an Australian Army officer currently serving as Aide-de-Camp to Commander Army Aviation Command. He holds a Bachelor of Science from the Australian Defence Force Academy, and his work has been published in a variety of fora, including the Australian Strategic Policy Institute's *The Strategist*, the Australian Defence College's *The Forge*, and *Grounded Curiosity*.

Peter Yule is a research fellow at the School of Historical and Philosophical Studies, the University of Melbourne. He has written widely on Australian military, economic, and medical history. His books include histories of the Collins Class submarine project, Australian National Airways, and the Australian economy in the First World War, and biographies of Sir James Rowland and business leaders W. L. Baillieu and Sir Ian Potter. His latest books are *The Long Shadow: Australia's Vietnam Veterans Since the War* (NewSouth Publishing, 2020) and *Vic Bar: A History of the Victorian Bar* (Australian Scholarly Publishing, 2021).

Foreword

As the inaugural Defence Space Commander, I take great delight in having this opportunity to introduce *Australian Perspectives on Global Air and Space Power: Past, Present and Future*. This book represents a milestone for Air and Space Power in Australia for a number of important reasons, all of which demonstrate how far we have come since the inhabitants of our wide continent first gazed to the skies, and beyond to the stars.

First, this volume draws together voices across the spectrum of air and space power stakeholders. The existing body of literature – regrettably scant at present – has tended to focus on drawing from one particular cohort, and so we have had individual works from the collected thoughts of the academic community, military practitioners (invariably very senior officers), and, more recently, junior military personnel in the form of contributions to any one of the number of military-focused online fora. This book stands apart from the rest as it seeks to draw these voices together in dialogue, rather than talking past each other, as is possible when conversations are compartmentalised. As such, this volume presents the varied perspectives of each author in something of a dialogue in which the voice of the sergeant shares equal footing with that of the air marshal.

Second, and as evident in the title, this book seeks to reinforce the formal linkage of air and space power. For many years, and despite its emergence as a fundamental enabler of virtually all military capability, space power remained an ethereal concept. As we have seen with cyberspace, the ubiquity and fundamental utility of space capabilities can complicate efforts to constrain them through the traditional military command and control relationships with which aviators are familiar. This challenge very much reflects one of the key questions of this book; namely, does technology drive social behaviour, or do our social structures (the military being one of the more *structured* social structures!) drive our appreciation and relationship with technology? The creation of Defence Space Command is an important step in recognising the need for specialist expertise in the space domain, a contemporary resonance of the turmoil of a century ago when aviators first agitated for a service of their own.

Finally, this book is a milestone in women's participation in the Air and Space Power dialogue in Australia, a field that has long been dominated by male voices. Of the 20 contributors to the volume, five are women, including two-thirds of the editing team. This figure sits just above the participation level of women in Air Force, which has grown steadily from around 18% in 2012. While it is easy to obsess over quantitative data such as this, I make this observation merely to indicate the fact that we are only today starting to meaningfully include women's voices in the literature that will form the basis of tomorrow's thinking on air and space power. As articulated in the landmark United Nations Security Council Resolution 1325 of 2000, the full, meaningful, and equal inclusion of women in all aspects of society is a fundamental requirement for international peace and security. Cultural change is hard, particularly in military organisations, but this book, and its diverse collection of authors, is an important indicator that we are moving closer to achieving broader inclusivity in our national security community.

Air Vice-Marshal Cath Roberts, AO, CSC
Commander Defence Space Command
Canberra, March 2023

Preface

As the Royal Australian Air Force moves into its second century, it is fitting to pause and reflect on how far we have come since our forbears' first tentative flights at Point Cook; flights that began an astonishing process of technological development that propelled us from aircraft made of fabric and wood, to the networked, globally responsive Air Force I am proud to lead today. History is only part of our story however, and we risk losing our competitive advantage by focusing too closely on where we have been. It is therefore important we develop the capacity in our people to critically analyse the contemporary strategic environment to inform today's decisions while also keeping an eye on the developments of tomorrow. In short, a strong Air Force needs people capable of synthesising yesterday's lessons in the context of today's geostrategic environment to inform action tomorrow.

I am proud, therefore, to present this volume that so ably reflects the diversity of perspectives contributing to the Australian air and space power dialogue. As the title suggests, in these pages, you will read of where we have been, the complex challenges we face today, and views on how both should inform our approach to defending Australia into the future. I am especially excited to see many contributions from serving Air Force members – that these aviators have taken the time to pen their contribution, over and above their primary role in Air Force gives me great comfort that the future of the organisation is in good hands. It is also heartening to see contributions from voices outside the military; our ability to prevail in the air and space domains depends on strong partnerships between government, academia, and industry, all of whom should be part of the conversation informing defence policy.

Finally, and most importantly, Australia has no air and space power capability without the Australian people who have, for over one hundred years, selflessly contributed in many ways to ensuring the safety and security of our nation. This book represents a unique way of engaging and including a broader audience than is traditionally involved in debates surrounding national defence. While we often reflect on the symbolic nature of Air Force as representative of the nation's values, we should also focus on the organisation as representative of the nation's *people*. A nation does itself a disservice by keeping its military separate from the society it serves, and it is my hope that this volume sparks interest in Australia's air and space power in new corners of society, thereby drawing us all closer as key contributors to the safety and security of future generations of Australians.

Air Marshal Rob Chipman, AM, CSC
Chief of Air Force
Canberra, March 2023

Disclaimer

 **AIR AND SPACE POWER CENTRE**

The views presented by the authors in this collection are the result of independent research projects. As such, they may not represent those of the Royal Australian Air Force, Australian Defence Force, nor the Australian Government.

Acknowledgements

This book grew from a series of seminars hosted at the University of New South Wales, Canberra (UNSW Canberra) between 2017 and 2019. A joint initiative between UNSW Canberra's Australian Centre for the Study of Armed Conflict and Society (ACSACS) and the Royal Australian Air Force's Air and Space Power Centre (formerly the Air Power Development Centre), the seminar series honoured the contribution of Air Marshal Sir James Rowland, the Chief of the Air Staff between 1975 and 1979. This edited collection is the product of these seminars, and its completion is a testament to the contributions, efforts, and perseverance of a great many people. It would not have eventuated had it not been for Professor Tom Frame, who, as Director of ACSACS, had the vision to eventually convert the collected Sir James Rowland Seminar papers into a publication. As such, the roots of this project can be traced to his early efforts and those of his colleagues and successors at ACSACS, specifically Andrew Blyth (ACSACS Manager, 2017), Professor Rob McLaughlin (Director ACSACS, 2018–2019), and Kus Pandey (Manager ACSACS, 2018–2019).

This endeavour has benefitted greatly from the generous support of the other major partner in the Rowland Seminar series, the Royal Australian Air Force. The series received strong support from the Air Force, with the Deputy Chief of Air Force, Air Vice-Marshal Warren McDonald, AM, CSC, launching the inaugural seminar in 2017. Chief of Air Force Fellow Wing Commander Dr Lewis Fredrickson and Group Captain Andrew Gilbert were also integral to the continuation of the seminar series. In particular, the Air and Space Power Centre remained steadfastly focused on publishing the collected papers, despite a relentless series of setbacks. No one expected a global pandemic to intervene, but the Air Force's continued support enabled us to bring this project to fruition, albeit much later than initially anticipated. Special thanks are owed to the Director of the Air and Space Power Centre, Group Captain Jason Baldock, who stretched his limited resources to support the publication process. In particular, Flight Lieutenant Karyn Markwell's keen copy-editing and Dr Vincent Daria's final manuscript review were essential to pulling together a cohesive book from the raft of seminar papers that were delivered over the course of the Rowland series.

Without the hard work of those who participated in the initial seminar series, this volume could not have been compiled. We thank all those who presented papers and attended each of the Sir James Rowland seminars, especially those who subsequently contributed to this edited collection. The many contributors to this volume have been a delight to work with, and they have been exceedingly accommodating as the closure of ACSACS, and then the global pandemic threatened to derail the project entirely. Special thanks are also owed to the Chief of Air Force's Air Power Fellows, Air Commodores Tony Forestier and Phil Champion. These gentlemen have been central to developing a cadre of Air Force members who can straddle the operational and academic while speaking with credibility in both worlds. Their patient support and wise counsel have helped raise the strategic acumen of the Air Force, and the large number of serving Air Force members among the authors in this book is a testament to the success of this endeavour.

In assessing this book's prospects, Routledge engaged several anonymous reviewers, to whom we owe a large debt of gratitude for giving generously of their time to provide detailed, constructive critique. The world of academic publishing heavily depends on the kindness of strangers who share their expertise and analysis with no reward other than the knowledge that they are contributing to furthering the discourse. Similarly, we want to thank Routledge for partnering with us to raise awareness of Australian perspectives on air and space power. We hope this volume will prompt others to add their voices to what is currently a small body of academic literature to help build a strong programme of multidisciplinary Air and Space power analysis in Australia.

Finally, we thank our families. Writing and editing books can be isolating, but the support they offer and the nights together they forgo to allow us to finish and submit the book are too often forgotten in the maelstrom of simply 'getting it done'. Special thanks are also owed to Dr Jordan Beavis, who assisted the editors in putting together the references and bibliographies for this collection.

Note on Terminology

The term used to refer to members of the Royal Australian Air Force (RAAF) has changed over time. Historically, the terms 'airman' or 'airmen' were used to denote members of the RAAF. As increasing numbers of people of various genders enter the Air Force, the gendered nature of the term has been considered, and in April 2021, then Chief of Air Force Air Marshal Mel Hupfeld announced the term would be replaced by 'aviator'. Although 'airman' and 'airmen' had been used to refer generally to all members of the Air Force, regardless of gender or rank, the introduction of this gender-neutral term reflected the RAAF's push to increase female representation in its ranks. Accordingly, this book uses the term 'aviators' in place of airmen. The exception is Chapter 2, which, as a historically based chapter, uses the relevant period's terminology. Similarly, 'uninhabited' or 'uncrewed' has been substituted for 'unmanned' where appropriate (e.g., uncrewed aerial vehicles rather than unmanned aerial vehicles).

Defence forces are infamous for their propensity to coin and use acronyms. In an effort to make this book accessible to general and non-specialist readers, the editors have endeavoured to remove acronyms where possible and appropriate. Where acronyms are used, terms are on first use provided in full, followed by the acronym in brackets. This process is repeated in every chapter.

Acronyms and Abbreviations

ACSACS	Australian Centre for the Study of Armed Conflict and Society
ACSC	Australian Command and Staff College
ADF	Australian Defence Force
AI	Artificial Intelligence
ANZAM	Australia-New-Zealand-Malaya
APDC	Air Power Development Centre
APM	The Air Power Manual
APU	Aircraft Performance Unit
ASA	Australian Space Agency
ASAT	Anti-satellite
ASPC	Air and Space Power Centre
ASPI	Australian Strategic Policy Institute
AUSMIN	Australia–United States Ministerial Consultations
AUSSpOC	Australian Space Operations Centre
AV	Air Vehicle
AWD	Air Warfare Destroyer
AWM	Australian War Memorial
BLOS	Beyond Line of Sight
CAAG	Cotton Aerodynamic Anti-G
CAC	Commonwealth Aircraft Corporation
CAF	Chief of Air Force
CAS	Chief of the Air Staff
CDF	Chief of the Defence Force
CO	Commanding Officer
CSpO	Combined Space Operation
CTs	Communist terrorists
CubeSat	Cube Satellite
DDE	Doctrine of Double Effect
DFC	Distinguished Flying Cross
DST	Defence Science and Technology
DVA	Department of Veterans' Affairs
ELDO	European Launcher Development Organisation
FAA	Fleet Air Arm

FAC	Forward Air Controller
FFS	Franks Flying Suit
FIC	Fundamental Inputs to Capability
FPRC	Flying Personnel Research Committee
GBAD	Ground-Based Air Defence
GCS	Ground Control Station
GFC	Global Financial Crisis
GPS	Global Positioning System
GWOT	Global War on Terror
HALE	High-Altitude Long-Endurance
HGV	Hypersonic Glide Vehicle
HIFIRE	Hypersonic International Flight Research Experimentation
IADS	Integrated Air Defence System
ICBM	Intercontinental Ballistic Missile
ICT	Information and Communications Technology
IED	Improvised Explosive Device
IoMT	Internet of Military Things
IoT	Internet of Things
ISF	Iraqi Security Forces
ISIS	Islamic State of Iraq and Syria
ISR	Intelligence, Surveillance, and Reconnaissance (ISR)
ISTAREW	Intelligence, Surveillance, Targeting, Acquisition, Reconnaissance and Electronic Warfare
IT	Information Technology
IWM	Imperial War Museum
LOS	Line of Sight
MALE	Medium-Altitude Long-Endurance
MTUAS	Maritime Tactical Unmanned Aerial Systems/Maritime Tactical Uncrewed Aerial Systems
NAA	National Archives of Australia
NASA	National Aeronautics and Space Administration
NEN	Near Earth Network
NGW	New-Generation Warfare
NHMRC	National Health and Medical Research Council
NLA	National Library of Australia
NUAS	Nano Unmanned Aerial Systems/Nano Uncrewed Aerial Systems
OODA	Observe–Orient–Decide–Act
PED	Processing, Exploitation, Dissemination
PLARF	People's Liberation Army Rocket Force
PMET	Professional Military Education and Training
PNT	Positioning, Navigation, and Timing
POW	Prisoner of War
RAAF	Royal Australian Air Force
RAF	Royal Air Force

RAN	Royal Australian Navy
RCAF	Royal Canadian Air Force
RFC	Royal Flying Corps
RN	Royal Navy
RNAS	Royal Naval Air Service
RPAS	Remotely Piloted Aircraft Systems
SABRE	Synergetic Air-Breathing Rocket Engine
SATCOM	Satellite Communication
SCIFiRE	Southern Cross Integrated Flight Research Experiment
Scramjet	Supersonic Combustion Ramjet
SDA	Space Domain Awareness
SMAINTSO	Senior Maintenance Staff Officer
SOCOM	Special Operations Command
SPU	Space Policy Unit
SSA	Space Situational Awareness
SUAS	Small Unmanned Aerial Systems/Small Uncrewed Aerial Systems
SVBIED	Suicide Vehicle-borne Improvised Explosive Device
TNA	The National Archives (UK)
TUAS	Tactical Unmanned Aerial Systems/Tactical Uncrewed Aerial Systems
UAS	Uninhabited/Uncrewed Aerial Systems
UAV	Uninhabited/Uncrewed Aerial Vehicles
UCLASS	Uncrewed Carrier-Launched Airborne Surveillance and Strike
UNCOPUOS	United Nations Committee on the Peaceful Uses of Outer Space
UNSW	University of New South Wales
USAAF	United States Army Air Force
USAF	United States Air Force
USAFA	United States Air Force Academy
USN	United States Navy
USO	United Service Organizations
WRE	Weapons Research Establishment
XO	Executive Officer

Introduction

An Australian perspective on air and space power

Nicole Townsend

When the *Wright Flyer* first lifted off the ground at Kitty Hawk, North Carolina, its inventors—two American brothers and aviation pioneers, Orville and Wilbur Wright—had no concept of the practical uses for their creation: the first motor-operated aeroplane. They had only one thing in mind—proving humans could fly. However, once the brothers established the viability of flight, they soon realised their technology could serve a greater purpose, specifically when used for military purposes. Certainly, the precedent for the military use of air assets had been set in 1794 when *la Compagnie d'aérostiers*, the French Aerostatic Corps (literally the Company of Balloonists), became the first air unit to employ air power through the use of balloons, including at the Battle of Fleurus.[1] Although the Wright brothers felt strongly enough that their aircraft could offer real potential, particularly if used for observation and reconnaissance, they were knocked back when they brought their ideas to the United States War Department. At the time, they thought their invention had the potential to prevent wars. The potential capabilities of aircraft could, they contended, make it 'so inadvisable that no government would dare to start' a war.[2] Still, like many new technologies, the aeroplane met with initial disdain from some quarters. French General Ferdinand Foch reputedly quipped in 1913—ten years after the *Wright Flyer* took to the sky—that although aviation as a sport was 'fine', it would be useless as an instrument of war. In reality, the aeroplane went on to revolutionise warfare by bringing it into the third dimension.[3]

More than a century later, air power has become an established and vital part of a military force's warfighting capability, providing not only tactical strike and close air support but also airlift capacity; intelligence, surveillance, and reconnaissance (ISR) capabilities; and deterrent effects, among other capabilities and roles. Yet the term 'air power' itself remains contested; indeed, Winston Churchill identified air power as 'the most difficult of all forms of military force to measure, or even express in precise terms'.[4] The ability of air forces to employ air power to achieve both tactical and strategic effects, combined with the breadth of air power capabilities now available and employed by modern air forces, perhaps explains the continued ambiguity in defining what exactly is meant by the term. At its heart, air power is, as William 'Billy' Mitchell famously stated, simply the 'ability to do something in the air'.[5] However, as Mark Clodfelter posits, such

DOI: 10.4324/9781003230656-1

simplified but succinct definitions are 'too vague', obfuscating the multiplicity of facets inherent in the term.[6] US Air Force Lieutenant Colonel Johnny R. Jones offers a more comprehensive definition:

> Air power is the integrated employment of all air and space forces to control and exploit the air and space environments to achieve national security objectives. Air power exploits the properties of its operating medium to realize unique operational characteristics and thus employ unique capabilities to provide the nation [with] a broad range of military options.[7]

However, despite the plethora of definitions that abound, a lack of consensus remains.

The emergence of the commercial and military exploitation of space as an eminent issue in recent years has only further muddied the waters.[8] The Soviet Union's success in launching the 84-kilogram Sputnik 1 satellite into a low earth orbit on 4 October 1957 marked the beginning of the 'Space Race' between the United States and the Soviet Union. It also marked the start of a journey that in the past 65 years has led to an increased focus on the commercial and military use of space. A little over a century since that first successful flight at Kitty Hawk, the boundaries of air power have expanded such that a new, fourth dimension of warfare has emerged: the space domain. Is the exploitation of space an expansion of air power, or is it more aptly considered as another, separate ability that is characterised by unique doctrines, strategies, capabilities, operations, and risks?

For the purposes of this book, air power is defined as the employment of all systems that can be used to exploit the third dimension of war—the air and space domain—either through hard or soft power. Included among these systems are those employed not only by air forces but also by other arms of the military forces that employ capabilities or systems that exploit the air domain. Reflecting recent developments in Australia that recognise the importance of the space domain to military operations and its integral influence on Australian national interests, this book presents the space domain and, in turn, space power, as separate to air power. The definition of space power adopted in this book is that expressed by the Royal Australian Air Force (RAAF) in the Space Power eManual, its first foundational reference for the employment of space power: 'Space power is the total strength of a nation's ability to conduct and influence activities to, in, through and from space to achieve its objectives'. Together, air and space power are two parts of military power that provide governments, including the Australian Government, with options to achieve strategic outcomes.[9]

An Australian Perspective

Australian military aviation began a decade after the *Wright Flyer* took to North Carolinian skies. At the 1911 Imperial Conference held in London, it was decided that the British Empire would develop aviation as a military capability. Two years later, the Commonwealth Military Forces stood up the Central Flying School at Point Cook in the south-eastern Australian state of Victoria. With the School's inception, Australia became the first within the British Empire to follow through on the decision made at the conference.[10] By the time war broke out in Europe in 1914, the nascent air force had become known as the Australian Flying Corps, the precursor to what would in 1921 become the RAAF.

Since its formation, the RAAF has been involved in numerous operations, both in Australia and internationally. Australian aviators have served in global conflicts, including both world wars, during the Cold War in the Korean and Vietnam Wars, the Malayan Emergency, and more recently, in Iraq, Afghanistan, and Syria. Although the RAAF is tasked with providing air and space power for military purposes and maintaining Australian security, these military operations have been interspersed with a variety of peacetime and humanitarian operations. For instance, RAAF assets most recently collaborated with the Australian Army and the Royal Australian Navy (RAN) on Operation Tonga Assist 2022, part of the Australian Government's support to the island nation of Tonga following the eruption of the Hunga Tonga-Hunga Ha'pai underwater volcano and the subsequent tsunami that struck the small archipelago on 15 January 2022.[11] Likewise, the RAAF assisted in the aftermath of the 2015 Nepal earthquake (including aeromedical evacuation and logistics support), aided in securing the wreckage site following the shooting down of Malaysian Airlines flight MH17 over Ukraine in 2014, and conducted evacuation flights following the Bali bombings in 2002 and 2005. However, one of its largest humanitarian responses followed the 2004 Boxing Day Tsunami and involved the provision of both Australian Defence Force (ADF) personnel and humanitarian aid to Sumatra.[12]

Air power has in recent years become a growing field of research that has seen the publication of myriad essays, articles, books, and other texts that encompass various aspects of the field and straddle a variety of disciplines and subdisciplines. John Andreas Olsen's *Global Air Power* and the *Routledge Handbook of Air Power* offer readers a fundamental understanding of air power. While the *Routledge Handbook of Air Power* provides readers with a grounding in the fundamentals of air power and all its aspects, from its foundations to its applications and strategic importance, *Global Air Power* interrogates how various global air forces have dealt with and employed air power.[13] These two books, together with numerous subject-based volumes published by the RAAF's Air and Space Power Centre (formerly the Air Power Development Centre), based in Canberra, Australia, allow us an understanding of the fundamentals of air power: what it is; how it began; how it is/has been employed; and how it has developed across different countries and regions of the world.[14] To these can be added John Olsen's *A History of Air Warfare*, which traces the use of air power throughout the 20th and into the 21st century, starting with the First World War and ending with the Second Lebanon War before considering the future path from air to space power.[15] Likewise, Jeremy Black's *Air Power: A Global History* provides a comprehensive and detailed examination of air power over land and sea.[16]

The literature on Australian air power is certainly not lacking, encompassing as it does biographies of key figures in the RAAF, squadron and unit histories that relate the operational experiences of Australian airmen and aviators since the First World War, and technologies and aircraft used in Australia.[17] To these may be added the various broader studies of the history of the RAAF and Australian air power.[18] Debates relating to Australian air power are also carried in online fora such as *Grounded Curiosity* and *The Cove*. Yet Australian experiences of the employment of air and space power still remain largely in the background of broader discussions of these dimensions of war. For instance, in Olsen's *Global Air Power*, Alan Stephens' chapter on air power in the Asia-Pacific region provides the only discussion of Australia's use of air power to underpin its national defence strategy.[19] Understandably, the overwhelming focus of the literature is not only smaller and middle powers such as Australia, but on the traditional and emerging 'great' powers, including the United States, Russia, the Soviet Union, and China. India and Israel are also emerging in the literature, as are commercial entities (such as SpaceX) and their actions in, influence on, and responsibility for the use of space. Few volumes provide a more holistic and encompassing overview of the challenges that have faced the RAAF throughout its existence, including those that are currently emerging and those that are yet to present themselves but may already be identified and considered. Fewer still draw on a broad spectrum of the current dialogue regarding air power, with many focusing instead on an individual aspect of air power.

Through a series of thematically grouped chapters, this book brings together established and emerging academics, professionals, and ADF officers from various disciplines to shine a light on Australian perspectives on the employment of air power and space power across the 20th and 21st centuries. As Air Vice-Marshal Cath Roberts notes in the Foreword to this volume, the editors have endeavoured to counter the echo chamber of air and space power dialogue that male voices have long dominated. Although there is certainly more work to be done in broadening women's participation in this and other such dialogues, the editors welcome the contribution of women to this volume. Likewise, a concerted effort has been made to include the voices of authors from across the spectrum of experience, not only those with an established position within their disciplines or discussions surrounding air and space power.

Adapted partly from papers presented at the Sir James Rowland Seminar Series, this book offers a primer for those seeking to understand historical and emerging issues within the realm of air power, covering aspects as diverse as the use of drones by non-state actors, the ethics of air strikes, the privatisation of air power, the historical trajectory of Australian air power strategy, and the sociological implications of an 'air force' identity.[20] Although many of its chapters use Australian-based case studies for their analysis, they offer broader applicability that speaks to issues and challenges inherent in and faced by other global air forces. The conflicts discussed herein include conventional and irregular wars across more than a century of armed conflict.

Structure

This book is divided into five thematic sections. Part One focuses on historical perspectives of air power and provides several close examinations of the employment of air power throughout the 20th century, demonstrating how the associated technologies and strategies have developed and changed and identifying lessons learned. In Chapter 1, Dr Michael Molkentin traces the evolution of the military use of aircraft in the Dominions of the British Empire before the First World War. Fast-forwarding to the Second World War, Dr Kristen Alexander addresses how Australian airmen held captive in Stalag Luft III—a Luftwaffe prisoner of war camp—continued to participate actively in the war despite their captivity. Contrary to their captors' taunts that for these prisoners, the war was over, Alexander demonstrates that as their fellow Australian airmen continued to serve across the world, including in Europe, the Mediterranean, the Middle East and the Pacific, Australian airmen held captive at this camp maintained a programme of 'active disruption' in the camp. In Chapter 3, Dr Peter Yule and Nicole Townsend consider the development of the strategic use of air power in Australia through the experience of a prominent Australian Chief of the Air Staff, Air Marshal Sir James Rowland. Dr Peter Hunter's chapter on the Malayan Emergency details the far-reaching consequences of the conflict on Australian air power strategy, suggesting that throughout the Cold War, Australia attempted to balance competing national security interests by establishing and maintaining independence in foreign and defence policy while also strengthening its alliances. Hunter concludes that these consequences and the lessons learned through counter-insurgency operations in Malaya were forgotten or 'unlearnt', as demonstrated in the subsequent Vietnam War.

The technology on which air forces depend is often considered at the expense of those people who create and use those technologies, however, the US Air Force suggested in 2013 that it was not aircraft but its people that represented the 'power behind the Air Force'.[21] The people who serve are as integral to air power as the aircraft they operate and the technologies that enable air forces. However, this acknowledgement also suggests a recognition that the social factors are as important, if not more important, than the 'toys' that draw the eye of many air power enthusiasts. Reflecting this, Part Two addresses the twin issues of identity and culture in the air force. Beginning with a Socratic dialogue, Air Commodore Jason Begley and Wing Commander Travis Hallen consider why air power needs to be the purview of an independent air force rather than an aviation wing or other such capability within the navy or army. Although the RAAF endured regular criticism and faced calls for its disestablishment and absorption into the Australian Army and RAN during the period between the two world wars, Begley and Hallen conclude that an independent air force is best placed to deliver air power because it possesses critical capabilities and expertise across air power, as well as a different perspective on the employment of air power. In Chapter 6, Dr Jarrod Pendlebury explores how identity has become a gatekeeper in Western air forces and how an air force identity is constructed and fostered within their cadets. Closing this section, Professor Tom Frame, in tracing the RAAF's attitude to education and training briefly, identifies a tension within the RAAF regarding education that has questioned, and perhaps underestimated, its importance in understanding the 'big picture', with training often considered pre-eminent.

Recognising the continual innovation of technology and its influence on the development and employment of air power capabilities, Part Three considers how some of these technological advancements have influenced the employment of air power. Dr Charles Vandepeer examines the Royal Air Force's Fighter Command, Bomber Command, and the invasion of Normandy in the Second World War, teasing out lessons for modern defence forces. Complementing Vandepeer, Dr Peter Layton's chapter on the privatisation of air power elucidates how the development of technology and the subsequent proliferation of cheaper and more accessible consumer aviation products has broadened the accessibility of air power capabilities beyond the usual states and entities. Where air power had been the domain of the state, providing those nations with an advantage over non-state actors, Layton illustrates how the growth of the consumer drone industry has opened the air domain to non-state actors such as the Islamic State of Iraq and Syria. Such non-state actors now use commercially available and inexpensive drones to facilitate a raft of capabilities, including reconnaissance, strike, and navigation. Squadron Leader Michael Spencer considers the impact of hypersonic air power on conventional air power, and Air Commodore Matt Hegarty discusses the relationship between organisational culture and technological innovation. Rounding out the section, Dr Peter Hobbins explores the influence anti-G suits had on the employment of air power in the Second World War.

Part Four steps into the 21st century, examining current issues faced by global air forces. Deane-Peter Baker considers the role of ethics in strategy formulation, and Group Captain Jo Brick argues for the consideration of a different conceptualisation of manoeuvre that seeks to unite the various elements of national power, including air power, in support of contemporary conflict. Rounding out the section, Amy Hestermann-Crane traces the development of Australia's use of the space domain, arguing that although Australia has made significant steps in its effort to ensure space security, there is still more to be done. The establishment of a whole-of-Defence space command, she argues, would allow a more concerted and separate focus on this domain and push Australia further towards becoming a contributing partner in building global space relationships.

The final section, Part Five, considers both emerging and anticipated problems that face not only global air forces such as the RAAF but also other military forces that employ air power, including the Australian Army. Lieutenant Colonel Keirin Joyce examines the concept of autonomy and considers the issues the Australian Army and the broader ADF face as they move towards becoming the most unmanned defence force with the continued development of Uncrewed or Uninhabited Aerial Systems (UAS) technologies. He questions whether the ADF is ready for such a transition. Officer Cadet James-Andre Galam contemplates the implications of military use of space for global defence forces and nations and the threats that stem from this new domain, both for the civilian and military spheres, focusing particularly on satellites their reliant technologies. Further, Lieutenant Christopher Wooding provides an overview of four key disruptive technologies modern militaries must confront, including cyber, space, UAS, and artificial intelligence.

Finally, although the editors of and contributors to this book endeavoured to contemporise its chapters as much as possible, it has not always been possible to maintain complete currency in light of the ever-evolving nature of the subject matter and the geostrategic developments that have emerged since the commencement of this project, including the global pandemic. Where possible, chapters have been updated to flag recent developments such as the stand-up of Defence Space Command in early 2022. Where this was not feasible, an editorial note has been included at the start of each affected chapter.

Notes

1 Nishant Gupta, *Indian Air Force in India's National Defence 2032* (New Delhi: KW Publishers Pvt Ltd in Association with Centre for Air Power Studies, 2014).
2 Peter L. Jakab, *The Published Writings of Wilbur and Orville Wright* (Washington, DC: Smithsonian Institution, 2000), 192.
3 John H. Morrow, Jr, "Expectation and Reality: The Great War in the Air," *Airpower Journal* (Winter 1996): 29, https://apps.dtic.mil/dtic/tr/fulltext/u2/a529474.pdf.
4 Winston S. Churchill, *The Second World War: The Gathering Storm*, vol. 1 (Boston: Houghton Mifflin Company, 1948), 100.
5 Johnny R. Jones, *William 'Billy' Mitchell's Air Power* (Honolulu: University Press of the Pacific, 2004), 1.
6 Mark Clodfelter, "Airpower Versus Asymmetric Enemies: A Framework for Evaluating Effectiveness," *Air & Space Power Journal* XVI, no. 3 (Fall 2002): 38.
7 Johnny R. Jones, "Air Power," *Air University* 1, www.airuniversity.af.edu/Portals/10/ASPJ/journals/Chronicles/jjones.pdf.
8 See for instance: Alec M. Robinson, "Distinguishing Space Power from Air Power: Implications for the Space Force Debate" (Thesis, Air University, 1998), https://spp.fas.org/eprint/98-239.pdf. Most nations, including Australia, incorporate space power within the remit of their air forces. The United States is an exception, having established the US Space Force as an independent branch of the US Armed Forces in 2019. The formerly independent Russian Space Forces were merged with the Russian Air Force to create the Russian Aerospace Forces in 2015, while China's People's Liberation Army Strategic Support Force is not exclusively a space force, incorporating cyber and electronic warfare capabilities. See: Matthew Bodner, "As Trump Pushes for Separate Space Force, Russia Moves Fast the Other Way," *DefenseNews*, 22 June 2018; Robert Farley, "Managing the Military Problem of Space: The Case of China, Part 2," *The Diplomat*, 24 May 2021.
9 Until 2022, space power was encompassed within the Royal Australian Air Force's Air Power Manual. In March 2022, the Australian Defence Force service chiefs launched the Space Power eManual alongside a revised edition of the Air Power Manual, in recognition of the role of space as a critical enabler of military capabilities and activities and the growing militarisation of space. See: Department of Defence, *Space Power eManual: Light-speed Edition* (Canberra: Commonwealth of Australia, 2022), www.airforce.gov.au/sites/default/files/doc/attachments/raaf-pages/213304_space_power_emanual_v1.0a.pdf.
10 South Africa, like Australia, was quick to act on the conference discussions in 1911, establishing the South African Aviation Corps in 1915 (though the first steps toward

its establishment had been taken in 1912). The Canadian Aviation Corps (the precursor to what is now known as the Canadian Air Force) was raised at the start of the First World War, but it comprised only three personnel and one aircraft and was short-lived, being stood down in May 1915. The Canadian Air Force was not formed until three years later, in 1918. New Zealand did not gazette a permanent air force until 1923, while the Indian Air Force did not stand-up until 1932. For more on British imperial air power during this period, see: Alex Spencer, *British Imperial Air Power: The Royal Air Forces and the Defense of Australia and New Zealand between the World Wars* (West Lafayette, Indiana: Purdue University Press, 2020).

11 Captain Zoe Griffyn, "ADF Supports Australia's Response to Tonga," *Defence News*, 31 January 2022.

12 For a brief outline of these and other humanitarian and disaster relief efforts conducted by the RAAF, see: Royal Australian Air Force, "Recent History of Air Force Humanitarian Assistance," *Air Force*, https://web.archive.org/web/20220321231432/www.airforce.gov.au/operations/humanitarian-support/recent-history-air-force-humanitarian-assistance. For a broader and in-depth analysis of Australian overseas emergency relief operations, including RAAF involvement, see: Steven Bullard, *In Their Time of Need: Australia's Overseas Emergency Relief Operations, 1918–2006* (Cambridge: Cambridge University Press, 2017).

13 John Andreas Olsen, ed., *Global Air Power* (Washington, DC: Potomac Books, 2011); John Andreas Olsen, *Routledge Handbook of Air Power* (Abingdon, Oxon: Routledge, 2018).

14 See for example: David Burns, ed., *Proceedings of the 2016 RAAF Air Power Conference: Multi-Domain Integration – Enabling Future Joint Success* (Canberra: Air Power Development Centre, 2017); Travis Hallen and Michael Spencer, *Hypersonic Air Power* (Canberra: Air Power Development Centre, 2018); Air Power Development Centre, *Air Power in a Disruptive World: Proceedings of the 2018 RAAF Air Power Conference* (Canberra: Air Power Development Centre, 2019); Don Woldhuis, *The Path to 5th Generation Electronic Warfare* (Canberra: Air Power Development Centre, 2019).

15 John Andreas Olsen, *A History of Air Warfare* (Washington, DC: Potomac Books, 2010).

16 Jeremy Black, *Air Power: A Global History* (New York: Rowman et Littlefield, 2016).

17 See, for instance: Alan Stephens and Jeff Isaacs, *High Fliers: Leaders of the Royal Australian Air Force* (Fairbairn, Australian Capital Territory: Air Power Development Centre, 1996); Mark Lax, *Alamein to the Alps: 454 Squadron RAAF, 1941–1945* (Wanniassa, Australian Capital Territory: Self-published, 2006); Adam Lunney, *Ready to Strike: The Spitfires and Australians of 453 (RAAF) Squadron Over Normandy* (Carlton, Victoria: Echo Books, 2018); Mark Lax, *From Controversy to Cutting Edge: A History of the F-111 in Australian Service* (Newport, New South Wales: Big Sky Publishing, 2021); Stewart Wilson, *Wirraway, Boomerang and CA-15 in Australian Service* (Weston Creek: Australian Capital Territory: Aerospace Publications, 1991).

18 See for example: Alan Stephens, *The Royal Australian Air Force: A History* (London: Oxford University Press, 2006); Alan Stephens, *Going Solo: The Royal Australian Air Force: 1946–1971* (Canberra: Australian Government Publishing Service, 1995).

19 Alan Stephens, "The Asia Pacific Region," in *Global Air Power*, ed. John Andreas Olsen (Washington, DC: Potomac Books, 2011), 299–333, EBSCOhost eBooks Collection.

20 The Sir James Rowland Seminar Series was a joint collaboration between the Royal Australian Air Force's Air and Space Power Centre (formerly the Air Power Development Centre) and the University of New South Wales' Australian Centre for the Study of Armed Conflict and Society. This book includes chapters that derive from seminar presentations delivered across these events, as well as the work of authors who did not participate in these seminars.

21 Mark A. Welsh III, "Global Vigilance, Global Reach, Global Power for America," *Air & Space Power Journal* (March–April 2014): 4, www.airuniversity.af.edu/Portals/10/ASPJ/journals/Volume-28_Issue-2/SLP-Welsh.pdf.

References

Air Power Development Centre. *Air Power in a Disruptive World: Proceedings of the 2018 RAAF Air Power Conference*. Canberra: Air Power Development Centre, 2019.

Black, Jeremy. *Air Power: A Global History*. New York: Rowman et Littlefield, 2016.

Bullard, Steven. *In Their Time of Need: Australia's Overseas Emergency Relief Operations, 1918–2006*. Cambridge: Cambridge University Press, 2017.

Burns, David, ed. *Proceedings of the 2016 RAAF Air Power Conference: Multi-Domain Integration – Enabling Future Joint Success*. Canberra: Air Power Development Centre, 2017.

Churchill, Winston S. *The Second World War: The Gathering Storm*, vol. 1. Boston: Houghton Mifflin Company, 1948.

Clodfelter, Mark. "Airpower Versus Asymmetric Enemies: A Framework for Evaluating Effectiveness." *Air & Space Power Journal* XVI, no. 3 (Fall 2002): 37–46.

Department of Defence. *Space Power eManual: Light-speed Edition*. Canberra: Commonwealth of Australia, 2022. www.airforce.gov.au/sites/default/files/doc/attachments/raaf-pages/213304_space_power_emanual_v1.0a.pdf.

Gupta, Nishant. *Indian Air Force in India's National Defence 2032*. New Delhi: KW Publishers Pvt Ltd in association with Centre for Air Power Studies, 2014.

Hallen, Travis, and Michael Spencer. *Hypersonic Air Power*. Canberra: Air Power Development Centre, 2018.

Jakab, Peter L. *The Published Writings of Wilbur and Orville Wright*. Washington, DC: Smithsonian Institution Washington, 2000.

Jones, Johnny R. "Air Power." *Air University*. www.airuniversity.af.edu/Portals/10/ASPJ/journals/Chronicles/jjones.pdf.

———. *William 'Billy' Mitchell's Air Power*. Honolulu: University Press of the Pacific, 2004.

Lax, Mark. *Alamein to the Alps: 454 Squadron RAAF, 1941–1945*. Wanniassa, Australian Capital Territory: Self-published, 2006.

———. *From Controversy to Cutting Edge: A History of the F-111 in Australian Service*. Newport, New South Wales: Big Sky Publishing, 2021.

Lunney, Adam. *Ready to Strike: The Spitfires and Australians of 453 (RAAF) Squadron Over Normandy*. Carlton, Victoria: Echo Books, 2018.

Morrow, Jr, John H. "Expectation and Reality: The Great War in the Air." *Airpower Journal* (Winter 1996): 27–34. https://apps.dtic.mil/dtic/tr/fulltext/u2/a529474.pdf.

Olsen, John Andreas. *A History of Air Warfare*. Washington, DC: Potomac Books, 2010.

———, ed. *Global Air Power*. Washington, DC: Potomac Books, 2011.

———. *Routledge Handbook of Air Power*. Abingdon, Oxon: Routledge, 2018.

Robinson, Alec M. "Distinguishing Space Power from Air Power: Implications for the Space Force Debate." Thesis, Air University, 1998. https://spp.fas.org/eprint/98-239.pdf.

Royal Australian Air Force. "Recent History of Air Force Humanitarian Assistance." *Air Force*. https://web.archive.org/web/20220321231432/www.airforce.gov.au/operations/humanitarian-support/recent-history-air-force-humanitarian-assistance.

Spencer, Alex. *British Imperial Air Power: The Royal Air Forces and the Defense of Australia and New Zealand between the World Wars*. West Lafayette, Indiana: Purdue University Press, 2020.

Stephens, Alan. *Going Solo: The Royal Australian Air Force: 1946–1971*. Canberra, Australian Government Publishing Service, 1995.

———. *The Royal Australian Air Force: A History*. London: Oxford University Press, 2006.

———. "The Asia Pacific Region." In *Global Air Power*, edited by John Andreas Olsen, 299–333. Washington, DC: Potomac Books, 2011. EBSCOhost eBooks Collection.

Stephens, Alan, and Jeff Isaacs. *High Fliers: Leaders of the Royal Australian Air Force*. Fairbairn, Australian Capital Territory: Air Power Development Centre, 1996.

Welsh III, Mark A. "Global Vigilance, Global Reach, Global Power for America." *Air & Space Power Journal* (March–April 2014): 4–10. www.airuniversity.af.edu/Portals/10/ASPJ/journals/Volume-28_Issue-2/SLP-Welsh.pdf.

Wilson, Stewart. *Wirraway, Boomerang and CA-15 in Australian Service*. Weston Creek, Australian Capital Territory: Aerospace Publications, 1991.

Woldhuis, Don. *The Path to 5th Generation Electronic Warfare*. Canberra: Air Power Development Centre, 2019.

Part One
Historical Perspectives

1 Military aviation in the British Dominions before the First World War

Michael Molkentin

The manner in which Britain's political and military authorities responded to the emergence and rapid development of aviation technology before the First World War is the subject of a substantial and well-researched body of scholarship.[1] However, the literature tends to focus on military aviation's evolution at the imperial centre; almost without exception, it gives no consideration to how Britain's colonial societies perceived and reacted to the new aeronautical inventions that emanated from Europe and the United States (US) during this period.[2] Histories of pre-war aviation in the dominions tend to focus on their subject in isolation from the other colonies and upon the role of individuals rather than institutions.[3]

This chapter considers pre-war military aeronautics in the dominions in a broader, comparative manner; its scope encompasses the military forces of Britain's self-governing dominions as well as the Anglo-Indian Army. It draws on research conducted in Australian, New Zealand, Canadian and British archives and the aviation-themed literature and journalism of the period to compare how colonial authorities perceived aviation before 1914, and identify the factors that influenced their engagement with this new and disruptive form of military technology. It begins with a brief contextual outline of the adoption of military air service by Britain's armed forces.

Within the British Army, interest in aviation extends back to the 1850s and 1860s when, like other armies in Europe and America, experiments conducted by technically minded individuals suggested the potential of balloons on the battlefield. As early as 1862, the Army's Ordnance Select Committee declared 'the subject' of balloon aviation 'no longer an experimental one' and recommended their adoption by the British War Office.[4] It took until 1878 for the Army to acquire a nascent flying capability. British soldiers used balloons in manoeuvres and campaigns during the following two decades, most notably during the Second Boer War (1899–1902). Official experimentation with powered aircraft commenced in 1906 but progressed slowly; at the time, there was some debate as to whether dirigible airships or aeroplanes were more suited to military use. The rapid evolution of aeronautical technology in Europe from 1908–09, the employment of aircraft in the Italo-Turkish War (1911–12) and Balkan Wars (1912–13) and the Mexican Revolution (starting in 1910), and the adoption of aircraft by other European armies, prompted the War Office to establish an aviation unit equipped with

DOI: 10.4324/9781003230656-3

both aeroplanes and balloons. The critical factor in this decision appears to have been a series of conferences by the Committee of Imperial Defence, between July and October 1910, that frankly acknowledged the threat German airships posed to Britain's security—a threat the Royal Navy's long-held supremacy could not counter. Initially a unit of the Royal Engineers, the British Army's nascent flying arm became a corps in its own right in April 1912, when the Royal Flying Corps (RFC) was established. The RFC comprised a military (Army) and naval (Royal Navy) wing and a joint flying school.[5] By the beginning of the First World War, several British firms, besides a government-sponsored aircraft factory, were designing and producing aeroplanes. The RFC and Royal Naval Air Service (RNAS, as the 'naval wing' had become known) had, split roughly between them, 113 aircraft and 2073 officers and men. Relative to the size of its army, Britain went to war in 1914 with the 'most aeronautically inclined army' of all the belligerents.[6]

These pre-war developments in aviation in Europe first came to colonial societies through print: that is, via the press and aviation-themed literature. Aircraft had been the subject of fiction and speculative nonfiction even before the Wright brothers demonstrated powered flight in 1903.[7] Widely read science-fiction writers such as Jules Verne and H G Wells imagined future conflicts in which aircraft acted as decisive weapons.[8] As aircraft became a practical reality in the early 20th century, novelists in the British world speculated on their ramifications for defence, including in an imperial context. For instance, in Herbert Strang's 1913 adventure novel *The Air Patrol*, an Imperial flying corps defeats a Mongolian invasion of India.[9] Similarly, in two of Rudyard Kipling's short stories, a global authority known as the 'Aerial Board of Control' uses air power to enforce a kind of *Pax Britannica* on the entire world.[10] Examples of this theme also come from fiction produced within the colonies. Published in Australia, a story titled *The Command of the Air* imagines how, in 1912, a Commonwealth Aviation Corps singlehandedly defeats a Japanese invasion force, while a similar plotline features in Raymond Longford's 1913 film *Australia Calls*.[11] Characteristically, these fictional speculations foresaw in aircraft the means to project considerable power over the vast spaces that separated the British colonies. They took for granted that the relatively large leaps aeronautical technology made before the war would continue.

Among the nonfiction titles that reported and speculated on the implications of the air age for colonial audiences, *Aerial Warfare*, published in 1909 by the British motoring journalist R P Hearne, was perhaps the most influential.[12] Hearne predicted that aircraft would provide a means of securing the empire's fringes more economically than armies or navies—an argument that colonial aviation advocates would employ when lobbying their governments. The editor of one British journal, *The Aeroplane*, agreed, but for different reasons: to him, the aeroplane symbolised Britain's racial superiority and 'would give the white population [in the colonies] much the same advantage which they once possessed when they alone owned modern rifles'.[13]

Besides publishing reviews and extracts of these texts, the colonial press also reported aeronautical developments in Europe and North America, such as Louis

Blériot's channel crossing and the large air show at Rheims in 1909 and, later, the employment of aircraft during the Balkan Wars and the Mexican Revolution. In the South Australian capital of Adelaide, the city's two largest daily newspapers together published more than 2000 articles on aviation and aviation-related subjects between December 1909 and December 1913.[14] Other metropolitan centres around the empire appear to have been similar. As with the literature, colonial press coverage of aviation during the pre-war period seems to have frequently associated the new technology with defence matters. Indeed, it would not be surprising if readers of this material conceived aircraft, first and foremost, as weapons of war.

It is difficult to evaluate the effect such textual representations had on people's thinking. Nevertheless, it seems reasonable to suggest they cultivated an awareness of, and perhaps even interest in, aerial matters (an attitude that, from the late 1920s, became known as airmindedness) in the dominions and inspired some popular advocacy for local aerial defences. One interesting indicator, which Brett Holman has analysed, is a spate of sightings of unidentified flying objects in Australia, Britain, Canada, New Zealand and South Africa between the 1890s and 1918.[15] It is perhaps noteworthy that witnesses sometimes associated these with a foreign threat; in Australia, for example, one witness reported the crew of a craft calling to him in a foreign language, while another described the object he saw as looking like 'a Japanese airship'.[16] Perhaps significantly, immediately before these sightings, the Australian press had reported airship trials by Japan's armed forces.

Another indicator of burgeoning airmindedness is the appearance of aviation defence lobby groups in Britain and the dominions, which had the self-proclaimed purpose of advocating for the development of colonial air capabilities. The Aerial League of the British Empire was formed in London in early 1909 to promote aviation at home and throughout the empire. Agents General from some of the dominions served on its executive committee, and it established provincial branches elsewhere in the empire, including in Australia and Canada. Other unassociated organisations, such as the Aero Club of New Zealand and the Aerial Experimental Association in Canada, provided aviation enthusiasts in the dominions with forums to exchange knowledge and from which they could lobby local authorities.[17]

The Aerial League's Australian branch ostensibly had more success than most of its counterparts. In 1909, it convinced the federal government to sponsor a competition to design and build 'a flying machine for military purposes in Australia'. The contest failed dismally; public donations failed to finance the balance of the prize, and none of the forty-five entries met the government's technically ambitious criteria.[18] With few exceptions, aviation experimentation in all dominions lagged well behind Britain, the main European powers and the US before 1914, while the colonial manufacture of aircraft remained practically non-existent. The colonies simply lacked the scientific and industrial infrastructure to make much headway in the highly technical and technological arena of powered flight. Things were not helped by the mass exodus of talented young men interested in aviation

to Europe, where they could learn to fly and find employment in the burgeoning aero industry. Perhaps recognising this, the London-based Imperial Air Fleet
Committee, an initiative of the Overseas Club (a gentlemen's club formed to promote imperial cohesion), raised funds to supply an aeroplane to each dominion for
the establishment of a flying school, something it proclaimed, 'for the purposes of
the defence of the Empire and its commerce as a whole'. The Committee provided
a single Blériot monoplane to New Zealand in June 1913, another to Australia the
following year, and one to South Africa in 1917.[19]

The representations of European aircraft manufacturers—who, in the lead up to
the First World War, perceived a potential market in the colonial military forces—
appear to have had a greater effect on official thinking than these aerial defence
lobby groups. A demonstration tour by the British and Colonial Aeroplane Co.
during 1911 seems to have been particularly influential in exporting aviation to
the colonies. Display flights in India and Australia provided local military authorities with a practical demonstration of what aircraft might achieve on operations.
In India, the company's aircraft participated in manoeuvres, impressing the Indian
Army's Chief of the General Staff, Major General Douglas Haig, to the extent
that, in May 1911—a year before the formation of the RFC—he wrote to the War
Office requesting advice on starting an Indian Army flying school and corps.[20]
After India, the British and Colonial Aeroplane Co. travelled to Australia, dem-
onstrating the Bristol Boxkite machine to officers of the Commonwealth Mili-
tary Forces.[21] Their favourable reports prompted the Australian defence minister,
George Pearce, to publicly commit to starting an aviation corps in Australia.[22] In
South Africa, too, the representations of another private enterprise, the African
Aviation Syndicate Ltd, convinced the Union Government to seek advice from
London on aerial matters and, subsequently, to contract this company to train
army aviators.[23]

Aviation historians have tended to overemphasise the historical agency of individuals at the expense of other factors. Nevertheless, it seems that the attitudes
and actions of influential individuals had the potential to decisively shape the
extent to which the British dominions engaged in aviation before and during the
war. Before the war, some soldiers in each of the dominions and India recognised
the need to acquire an aerial capability like their counterparts in Britain. However,
the decision to invest in this novel and expensive technology typically came down
to the politicians and the government ministers responsible for their respective
dominion's defence/military portfolio.

Both before and during the First World War, George Pearce had a well-
documented enthusiasm for aviation and technological innovation more generally.[24] Whereas the Australian military establishment generally treated aviation
ambivalently before the war and, at best, wanted to wait and follow London's lead,
Pearce pressed ahead, making decisions to give Australia an air capability that, in
some instances, preceded developments in London. Indeed, Pearce publicly com-
mitted to raising an Australian 'aviation corps' in March 1911 (a month before
the British Army established its Air Battalion); later that year, he rejected the War
Office's advice to send soldiers to Britain to learn to fly, planning instead to open

a flying school in Australia; and, despite advice to the contrary, Pearce's department purchased its aeroplanes four months before the British Army held trials to determine which type of aircraft its flying corps would adopt.[25] On the other hand, Pearce's counterpart in Canada, the eccentric Sam Hughes, derided the emerging technology. By one account, Hughes regarded aeroplanes as 'costly toys, only as yet in the experimental stage'. Another account has him in 1914 dismissing aircraft as 'the invention of the Devil' and doubting they would play any role in 'such a serious business as the defence of the nation'.[26] Hughes's scepticism explains at least in part why, despite the relatively sophisticated and successful experimental work done by Canadian pioneers and two recommendations by the Canadian general staff to start a flying corps, the Canadian Government did less to develop aerial defences than some of its counterparts in the other dominions before the war.

Even so, colonial defence leaders such as Hughes and Pearce worked within political and strategic contexts that influenced their aviation policies. Foremost is how each of the colonial governments engaged with the concept of imperial defence. This strategic plan, developed at a series of Imperial Conferences in the years preceding the First World War, saw the self-governing dominions agree to take greater responsibility for the empire's defence by raising and maintaining local forces. These would allow dominions to provide for their regional protection and supply expeditionary forces to support Britain in the event of a major conflict. In return, the dominions—which in most cases agreed to organise and equip their forces in uniformity with London—would have access to British expertise and equipment.[27] For dominion military forces interested in acquiring an aviation capability before and during the war, this access to British industrial and technical expertise would prove critical. Indeed, all official discussions about military aviation in the dominions relied on advice from the War Office.

Each of the dominion governments weighed up the costs and benefits of imperial defence before planning their defence forces according to their own perceived strategic circumstances. For example, Australian and South African authorities perceived the colonial aspirations of Japan and Germany as regional threats, and they pursued the development of local military capabilities more vigorously than their Canadian counterparts, for whom the US' Monroe Doctrine provided a measure of security.[28] Ultimately, funding determined the extent to which the dominion governments acquired a military flying capability before the war. In the context of imperial defence, all increased their defence estimates substantially before 1914. However, the available funds needed to be shared between substantial defence projects such as compulsory military training schemes and the acquisition of naval vessels and infrastructure.

The Australian and Indian Governments invested in establishing military flying schools and aviation corps before the First World War. In both cases, however, aeronautics received only a tiny proportion of the defence budget: just 0.14 per cent of defence estimates between 1911 and 1914 in Australia's case.[29] The New Zealand general staff expressed a genuine interest in aeronautics and sent officers overseas to train with the RFC but deferred plans when the expense of the enterprise became clear.[30] The cost likewise represented a deterrent for South African authorities, though they changed their mind in 1912 when confronted with the prospect of German military aircraft in neighbouring South West Africa.[31] In 1913, the Union Defence Force began training ten officers at a private aviation school in South Africa. The War Office willingly provided these dominions with advice on establishing their aviation corps and invited officers from colonial armies to observe and train with the RFC in Britain. London's assistance proved crucial to the training of several colonial airmen before the war and the Australian and Indian Governments' purchase of equipment and employment of instructors for their schools. However, at no point did British authorities compel the dominions to make a start in flying. Indeed, London tended to emphasise the technology's nascent and rapidly evolving nature and recommended a circumspect approach.[32]

The First World War turned these tentative steps towards establishing a military flying capability in the dominion defence forces into a headlong rush forward. Australia, India and South Africa fielded units on operations between 1914 and 1918, and all dominions contributed personnel to Britain's flying services. Indeed, by the war's end, at least one in five of those killed while flying in the British flying services hailed from the dominions and colonies.[33] Ironically, given the reluctance of its defence leaders before the war, Canada's contribution far outstripped the others—more than 20,000 Canadians served in the Royal Air Force and its antecedents during the war.

As Canada's substantial contribution indicates, pre-war planning by colonial governments did not prove a prerequisite for contributing to the empire's war effort in the air. Indeed, Australian, Indian and South African efforts to establish their own, distinctly national contribution based on their pre-war plans proved, in many cases, impractical. The handful of colonial flying units that saw active service were unable to sustain themselves in the field and came to rely wholly on the RFC for equipment and, in some cases, personnel. Meanwhile, the colonies' government and private aviation schools struggled when British authorities refused to release aircraft and instructors to service them.[34] These were issues that writers, aviation lobbyists and many policymakers failed to foresee.

Through the experience of the First World War, therefore, colonial defence authorities would begin to understand the challenges inherent in raising and maintaining air services on a small scale. In confronting these challenges after 1918, the influences that had shaped aviation during the pre-war years continued to influence air leaders in the dominions. In the interwar period, air power throughout the empire evolved at the confluence of the dominions' respective strategic contexts, the reciprocal dynamics of imperial defence and a growing popular and professional recognition of the aeroplane's military value.

Notes

1 See for example, Alfred Gollin, *The Impact of Air Power on the British People and Their Government, 1909–14* (London: Macmillan in association with King's College, 1989); Alfred Gollin, "The Wright Brothers and the British Authorities, 1902–1909," *The English Historical Review* 95, no. 375 (April 1980): 293–320; Hugh Driver,

The Birth of Military Aviation: Britain, 1903–1914 (Suffolk: Boydell Press, 1997); Michael Paris, *Winged Warfare: The Literature and Theory of Aerial Warfare in Britain, 1859–1917* (Manchester: Manchester University Press, 1992); Peter Mead, *The Eye in the Air* (London: HMSO, 1983); Richard P. Hallion, *Taking Flight: Inventing the Aerial Age from Antiquity to the First World War* (Oxford: Oxford University Press, 2003); Charles Gibbs-Smith, *Aviation: An Historical Survey from Its Origins to the End of World War II* (London: HMSO, 1970); David Edgerton, *England and the Aeroplane: An Essay on a Militant and Technological Nation* (London: Macmillan, 1991).

2 For a notable exception see Michael Paris, "Air Power and Imperial Defence 1880–1919," *Journal of Contemporary History* 24, no. 2 (1989): 209–25. Since the presentation of this paper the author has also published an article that briefly considers the development of air power in British colonial military forces: Michael Molkentin, "The Dominion of the Air: The Imperial Dimension of Britain's War in the Air, 1914–1918," *British Journal of Military History* 4, no. 2 (2018): 70–90.

3 Histories of air power in the dominions include: F. M. Cutlack, *The Official History of Australia in the War of 1914–1918: Volume VIII: The Australian Flying Corps in the Western and Eastern Theatres of War 1914–1918* (Sydney: Angus and Robertson, 1923); Michael Molkentin, *Fire in the Sky: The Australian Flying Corps in the First World War* (Crows Nest, NSW: Allen and Unwin, 2010); Michael Molkentin, *Centenary History of Australia and the Great War, Volume 1: Australia and the War in the Air* (Melbourne: Oxford University Press, 2014); S. F. Wise, *Canadian Airmen and the First World War: Official History of the Royal Canadian Air Force, Volume I* (Toronto: University of Toronto Press, 1980); C.W. Hunt, *Dancing in the Sky: The Royal Flying Corps in Canada* (Toronto: Dundurn Press, 2009); Adam Classen, *Fearless: The Extraordinary Untold Story of New Zealand's Great War Airmen* (Massey: Massey University Press, 2017); Dick Silberbauer, "Origins of South African Military Aviation," *Cross & Cockade Journal (Great Britain)* 7, no. 4 (1976), 174–80.

4 Report, Ordnance Select Committee, 7 February 1862, The National Archives (UK) (TNA), AIR1/2404/303/1.

5 By the commencement of the First World War these two wings would have effectively split into two separate services: the Royal Flying Corps and the Royal Naval Air Service.

6 Edgerton, *England and the Aeroplane*, 16.

7 Paris, *Winged Warfare*, 15–64.

8 Ibid., 16.

9 Herbert Strang, *The Air Patrol: A Story of the North-West Frontier* (London: Hodder and Stoughton, 1913).

10 Rudyard Kipling, "With the Night Mail," *The Windsor Magazine*, December 1905; Rudyard Kipling, "As Easy as ABC," *The London Magazine*, 1912.

11 Lawrence Zeal, "Command of the Air: The Story of How an Australian Aeroplane Met the Japanese in the New Warfare," *The Lone Hand*, March 1911.

12 R. P. Hearne, *Aerial Warfare* (London: John Lane, The Bodley Head, 1909).

13 "Aviation and the Empire," *The Aeroplane*, 15 August 1912.

14 Data from TROVE Newspapers, <http://trove.nla.gov.au/newspaper>, based on a search of articles in *The Register* and *The Advertiser* with aviation-themed terms on October 30, 2017.

15 This research is presented in several posts on Holman's blog "Airminded," https:// airminded.org/. It is also presented in: Brett Holman, "The Enemy at the Gates: The 1918 Mystery Aeroplane Panic in Australia and New Zealand," in *Australia and the Great War: Identity, Memory and Mythology*, eds. Michael J. K. Walsh and Andrekos Varnava (Carlton, Victoria: Melbourne University Press, 2016), 71–96.

16 "The Mysterious Lights," *The Mercury*, 19 August 1909, 2; "Phantom Airships," *Kalgoorlie Western Argus*, 24 August 1909, 34; "Japan's Airship Secrets," *The Argus,*

24 July 1909, 7; "Japan's Airship Secrets," *Morning Bulletin*, 4 August 1909, 5; "Japanese Airships," *The Queenslander*, 7 August 1909, 38; "Japan's Airship Secrets," *The Dubbo Liberal and Macquarie Advocate*, 14 August 1909, 8.

17 Errol Martyn, *A Passion for Flight: New Zealand Aviation before the Great War, Volume 2: Aero Clubs, Aeroplanes, Aviators and Aeronaughts, 1910–1914* (Christchurch: Volpane Press, 2013), 76–78.

18 For a more detailed discussion on this competition see Molkentin, *Australia and the War in the Air*, 2–5.

19 "London Presents an Aeroplane to South Africa," *Flight*, 10 May 1917, 460.

20 Paris, "Air Power and Imperial Defence 1880–1919," 217–19.

21 Reports, Irving and Foot, 16 January 1911 & 27 February 1911, National Archives of Australia (NAA), A289, 1849/8/31.

22 "An Aviation Corps," *The Examiner (Launceston)*, 4 March 1911, 4.

23 Memorandum, South African High Commissioner to the Secretary, War Office, 17 May 1912, TNA AIR1/117/15/40/23.

24 John Connor, *Anzac and the Empire: George Foster Pearce and the Foundations of Australian Defence* (Melbourne: Cambridge University Press, 2011), 42–43.

25 Molkentin, *Australia and the War in the Air*, pp. 10–12.

26 Wise, *Canadian Airmen and the First World War: Official History of the Royal Canadian Air Force, Volume I*, 16–24.

27 Chris Coulthard-Clark, "Australian Defence: Perceptions and Policies, 1871–1919," in *The German Empire and Britain's Pacific Dominions: Essays on the Role of Australia and New Zealand in World Politics in the Age of Imperialism*, eds. Christopher Pugsley and John A. Moses (Claremont: Regina Books, 2000), 171.

28 Neville Meaney, *A History of Australian Defence and Foreign Policy: Volume I* (Sydney: Sydney University Press, 1976), 158, 165, 213; James Woods, *Militia Myths: Ideas of the Canadian Citizen Soldier* (Vancouver: University of British Columbia, 2010), 195; Glen St. J. Barclay, *The Empire Is Marching: A Study of the Military Effort of the British Empire 1800–1945* (London: Weidenfeld and Nicolson, 1976), 43–45; Richard A. Preston and Ian Wards, "Military and Defence Development in Canada, Australia and New Zealand: A Three Way Comparison," *War & Society* 5, no. 1 (May 1987): 9–10.

29 Meaney, *A History of Australian Defence and Foreign Policy: Volume I*, 277.

30 Appendices to the Journals of the House of Representatives, 1914 Session I, H-19, 20 June 1913 to 25 June 1914, New Zealand National Library.

31 Minute, Louis Botha 16 July 1912, TNA, AIR1/117/15/40/23.

32 See, for example: Cablegram, High Commissioner's office to Department of Defence, 1 March 1912, NAA, A2023, A38/3/221.

33 For an analysis of the dominion contribution to the air war see Molkentin, "The Dominion of the Air', passim.

34 For a detailed development of this argument regarding Australia's contribution, see Molkentin, *Australia and the War in the Air*, passim.

References

Archival Sources

National Archives of Australia

A2023: Department of Defence, Correspondence files, 1907–1917.
A289: Department of Defence, Correspondence files, 1894–1917.
The National Archives (UK).
AIR 1: Air Ministry: Air Historical Branch: Papers (Series I).

Government Publications

Appendices to the Journals of the House of Representatives (Wellington, New Zealand).

Published Sources

Barclay, Glen St. J. *The Empire is Marching: A Study of the Military Effort of the British Empire 1800–1945*. London: Weidenfeld and Nicolson, 1976.

Classen, Adam. *Fearless: The Extraordinary Untold Story of New Zealand's Great War Airmen*. Massey: Massey University Press, 2017.

Connor, John. *Anzac and Empire: George Foster Pearce and the Foundations of Australian Defence*. Melbourne: Cambridge University Press, 2011.

Coulthard–Clark, Chris. "Australian Defence: Perceptions and Policies, 1871–1919." In *The German Empire and Britain's Pacific Dominions: Essays on the Role of Australia and New Zealand in World Politics in the Age of Imperialism*, edited by Christopher Pugsley and John A. Moses, 155–72. Claremont: Regina Books, 2000.

Cutlack, F. M. *The Official History of Australia in the War of 1914–1918: Volume VIII: The Australian Flying Corps in the Western and Eastern Theatres of War 1914–1918*. Sydney: Angus and Robertson, 1923.

Driver, Hugh. *The Birth of Military Aviation: Britain, 1903–1914*. Suffolk: Boydell Press, 1997.

Edgerton, David. *England and the Aeroplane: An Essay on a Militant and Technological Nation*. London: Macmillan, 1991.

Gibbs–Smith, Charles. *Aviation: An Historical Survey from its Origins to the End of World War II*. London: HMSO, 1970.

Gollin, Alfred. "The Wright Brothers and the British Authorities, 1902–1909." *The English Historical Review* 95, no. 375 (April 1980): 293–320.

———. *The Impact of Air Power on the British People and Their Government, 1909–14*. London: Macmillan in association with King's College, 1989.

Hallion, Richard P. *Taking Flight: Inventing the Aerial Age from Antiquity to the First World War*. Oxford: Oxford University Press, 2003.

Hearne, R. P. *Aerial Warfare*. London: John Lane, The Bodley Head, 1909.

Holman, Brett. "The Enemy at the Gates: The 1918 Mystery Aeroplane Panic in Australia and New Zealand." In *Australia and the Great War: Identity, Memory and Mythology*, edited by Michael J. K. Walsh and Anfreko Varnava, 71–96. Carlton, Victoria: Melbourne University Press, 2016.

Hunt, C. W. *Dancing in the Sky: The Royal Flying Corps in Canada*. Toronto: Dundurn Press, 2009.

Martyn, Errol. *A Passion for Flight: New Zealand Aviation before the Great War, Volume 2: Aero Clubs, Aeroplanes, Aviators and Aeronaughts, 1910–1914*. Christchurch: Volpane Press, 2013.

Mead, Peter. *The Eye in the Air*. London: HMSO, 1983.

Meaney, Neville. *A History of Australian Defence and Foreign Policy, Volume I*. Sydney: Sydney University Press, 1976.

Molkentin, Michael. *Fire in the Sky: The Australian Flying Corps in the First World War*. Crows Nest, NSW: Allen and Unwin, 2010.

———. *Centenary History of Australia and the Great War, Volume 1: Australia and the War in the Air*. Melbourne: Oxford University Press, 2014.

———. "The Dominion of the Air: The Imperial Dimension of Britain's War in the Air, 1914–1918." *British Journal of Military History* 4, no. 2 (2018): 70–90.

Paris, Michael. "Air Power and Imperial Defence 1880–1919." *Journal of Contemporary History* 24, no. 2 (1989): 209–25.

———. *Winged Warfare: The Literature and Theory of Aerial Warfare in Britain, 1859–1917*. Manchester: Manchester University Press, 1992.

Preston, Richard A., and Ian Wards. "Military and Defence Development in Canada, Australia and New Zealand: A Three-Way Comparison." *War & Society* 5, no. 1 (May 1987): 1–22.

Silberbauer, Dick. "Origins of South African Military Aviation." *Cross & Cockade Journal (Great Britain)* 7, no. 4 (1976): 174–80.

Strang, Herbert. *The Air Patrol: A Story of the North–West Frontier*. London: Hodder and Stoughton, 1913.

Wise, S. F. *Canadian Airmen and the First World War: Official History of the Royal Canadian Air Force, Volume I*. Toronto: University of Toronto Press, 1980.

Woods, James. *Militia Myths: Ideas of the Canadian Citizen Soldier*. Vancouver: University of British Columbia, 2010.

2 The Australian prisoner of war experience in Stalag Luft III, 1942–45[1]

Kristen Alexander

Air power underpinned Allied victory in the Second World War. Australian airmen in operations rooms, aircraft, and ground-support roles played a considerable part in achieving tactical and strategic success. If captured, however, the operational airman was generally perceived as being *hors de combat*, with seemingly no place in air power operations. But that was not the case. This chapter discusses how the Australian airmen[2] held as prisoners of war (POWs) in Stalag Luft III—a Luftwaffe POW camp located near Sagan (now Żagań, west Poland) in the German province of Lower Silesia, most notable for the so-called Great Escape— energetically continued to fulfil their service duties despite captivity. Drawing on official records as well as wartime writings, post-war interviews, memoirs, and medical testimony, it highlights how, rather than giving in to what one man termed the 'futility of the existence',[3] the Australian airmen overcame the challenges of captivity by enacting Royal Air Force (RAF) discipline and instituting a programme of active disruption that featured: small acts of defiance to maintain self-respect, active resistance, and theft; and the grand gestures of major escape plans, escape-related work, and escape.

Active disruption

351 Australian airmen spent time in Stalag Luft III's British compounds during its almost three years' operation as a POW camp.[4] Most had been in combat before their capture; they had bailed out of, or crawled from, crashed, burning, or sinking aircraft. After they were taken into custody, they were interrogated and sent to permanent POW facilities. Stalag Luft III was just one of the Luftwaffe's internment facilities. Before it opened in April 1942, Allied airmen had been incarcerated in Wehrmacht camps. After arriving at Sagan, British—including Australian— airmen were housed in four of the camp's six compounds, namely North, East, Belaria, and, for a time, Centre Compounds. Many captured airmen spoke of the 'shock' of captivity, the shame they felt on capture, and their abhorrence of the stigma associated with being a POW.[5] Their sense of disgrace was exacerbated when they heard the 'usual taunt' from their German captors: '*für sie der Krieg is beendet*'—'for you, the war is over'.[6]

DOI: 10.4324/9781003230656-4

Captivity was a serious blow to self-esteem and service pride. But rather than accept a situation defined by passivity and docility, the airmen rejected it. Accordingly, the airmen redefined the disruptive world of their Luftwaffe prison camp. They did away with the German designation of their captive state— *Kriegsgefangener* (war prisoner)—because of the negative and ignominious connotations that surrounded the word 'prisoner' and its associated trappings of captivity, such as POW number, fingerprinting, and identity discs. Instead, they collectively adopted the easier-to-pronounce abbreviation of 'Kriegie', derived from the first syllable—*krieg*—the German word for 'war'. By reappropriating and adapting the term, they removed the sense of derogation surrounding captivity and turned an afront into a linguistic badge of inclusiveness and pride. 'Kriegie' declared that they were still men of war, on operational service, enacting air power behind barbed wire,[7] and it became their preferred nomenclature during captivity and after the war.[8]

Demonstrating collective agency, the Kriegies institutionalised air force discipline in the camp. They acknowledged that running Stalag Luft III along RAF station lines was the best way to create order, and it was generally agreed that the senior officers 'did an excellent job'. As well as maintaining morale and discipline, they experienced a sense of security in adhering to the familiar aspects of their former service lives. Moreover, it afforded the airmen the support they needed to act disruptively. All new arrivals were interviewed by the senior British officer, who reminded the newly arrived Kriegies of their continuing service obligations. 'Those words', recalled one man, were 'clearly intended to galvanise us out of any apathy and meek resignation to our fate'.[9] Recalling the oft-repeated German refrain, he clearly remembered the group captain announcing, 'as though giving an order', that, 'For you . . . the war is *not* over'.[10] Indeed, 'it was our duty to make the enemy's task of imprisoning us as difficult and demanding on their resources as we could'.[11] By disrupting their captors, the Kriegies felt they could continue their contribution to the air war, albeit from behind barbed wire rather than in the skies above.

The airmen welcomed their new responsibility. Revealing their ingenuity and technical expertise, some constructed, maintained, and stole parts for radios, and others distributed transcripts from BBC news broadcasts to boost morale.[12] Parodying the Ten Commandments, an unknown wit penned the 'Kriegie's Commandments', a poem which, among other things, exhorted them 'to do no *arbeit*' (work) and to 'get into as many rackets as possible'.[13] At every opportunity, the men purloined German tools and supplies and suborned or bribed guards.[14] Other disruptive acts included 'the fierce joy of goon-baiting', which involved provoking the German prison staff.[15] One former POW recalled that they would 'harangue' and 'belittle' their captors. Whenever possible, they declared '*Deutschland kaput*' (Germany finished, or dead).[16] *Scangriff*, North Compound's gossipy news sheet, undermined German superiority: '[W]e always used to put a lot of garbage in it, so they never knew quite what was true', recalled one of the newsletter's contributors.[17] They relished their 'non-cooperative and aggressive' stance, and roll call was accordingly 'disorderly and covertly defiant', with many misbehaving and deliberately frustrating attempts to count their numbers.[18] Smoking on parade especially annoyed the Germans as it was considered *unsoldatlich*, that is, unsoldierly behaviour.[19]

Their disruptiveness did not go unpunished. Roll calls took longer, and the airmen were made to stand on parade for hours in inclement weather or long into the night.[20] They were put in the cooler—the prison cells—or confined to barracks, and Red Cross parcels and mail were withheld.[21] Some Kriegies suffered violence, with one recalling that he 'was hit by a German Rifle butt on the spine and was completely paralysed for two days'. He was 'threatened with death' on one occasion.[22] Intimidations, injuries, and collective sanctions did little to dent their ardour. Rather than *unsoldatlich* behaviour, the Kriegies believed their commitment to active disruption reflected their continuing service and air force pride.

The most significant aspect of active disruption was escape. As John Herington, the official historian of the Royal Australian Air Force's (RAAF) role in the air war against Germany and Italy, noted, all airmen knew they remained combatants despite their capture: 'they had a moral right and duty to escape if they could'.[23] This duty was underpinned by the Air Force Act and *King's Regulations* and formalised in the June 1941 edition of *Air Publication 1548*, which stressed that opportunities for escape would arise. It was an airman's duty to attempt to escape as, even if they failed, such attempts 'have a very appreciable nuisance value'.[24]

Each compound had its version of the escape or X organisation, as they called it. Australians participated in almost every aspect of its work. They drew up rosters disguised as participant lists for sports days, and sporty types created diversions.[25] The carpentry departments commandeered bed boards to shore up tunnels and built cabinets and hidey-holes to stow secret equipment.[26] Most of the Kriegies joined the army of 'watchers' (known as 'stooges'), who looked for any sign of the Germans,[27] while others carried out covert operations. Some meticulously constructed compasses or did metalwork, and others fabricated radio parts.[28] Photographers took passport photographs, and forgers replicated passes and identification papers.[29] Tunnellers dug, others carted dirt away, and gardeners disposed of it.[30] The tunnellers were so industrious that, in East Compound alone, between 60 and 70 tunnels were started during the first six months.[31]

Despite such diligent X work and prolific excavation, East Compound's only successful getaway occurred in October 1943. Dubbed 'the Wooden Horse' effort, three men made a 'home run' to England. None of the escapees were Australian, but Australian airmen had several important roles, including fabricating the horse, vaulting over it, and dispersing soil.[32] One played the harmonica to distract the guards.[33] Another slept in the bed of one of the escapees to cover up his absence on the night of the prison break.[34] If the escape route had not been discovered, another Australian airman would have been in the next batch that hoped to use the tunnel.[35] The culmination of North Compound's X work was the mass breakout of March 1944, in which six Australians were among the seventy-six who exited the camp. One was still in the tunnel when the attempt was discovered, and others awaited their chance. Only three made it back to England, with seventy-three recaptured. Their efforts did not go unpunished: 50 were executed in the post-escape reprisals, five of whom were Australian.[36] The duty to escape was rescinded in the breakout's aftermath, but an implied duty was still urged by encouraging alertness to escape.[37] Indeed, the Kriegies could do much to continue harassing the Germans, and they learnt 'very quickly' how they could continue their active disruption behind barbed wire, albeit on a small, highly localised scale.[38]

Evaluation

In June 1945, one newly liberated former Kriegie declared near-total commitment to escape among the Australians, most of whom had tried to escape at some point.[39] However, although this man's claim is a worthy testament to the Australians' involvement in active disruption, his assessment does not reflect reality. After liberation, Stalag Luft III's former Kriegies were asked to detail their escape work.[40] Although some of these debriefs have been lost, of those available, seventy-two were silent on their escape efforts. Ten per cent specifically asserted that they had made no escape bids. Both groups include airmen whose breakouts were recorded in other official reports or private accounts.[41] These contradictions suggest that the airmen interpreted escape in its narrowest form: exiting the camp. Had they been queried regarding the full spectrum of escape-related work—such as digging tunnels, forgery, or diversions so others could escape—the degree of their communal, if not individual escape-mindedness, and the extent of their disruptive acts would, perhaps, be more evident.

Reporting deficiencies aside, 100 per cent commitment can be nothing but an exaggeration, as not every POW could attempt to escape. For example, 21 per cent of the Australian airmen were captured after the March 1944 mass breakout; five per cent of those had been taken after the duty to escape had been rescinded in the wake of the Germans' decree that anyone caught in 'strictly forbidden' areas, which were classed as designated 'death zones', would be 'immediately shot on sight'.[42] Moreover, many men had been seriously wounded during the final moments of operations or after crashing or baling out of their aircraft. They were simply not physically capable of escaping. Additionally, the shock of capture engendered paralysing apathy in some men, which rendered them unable to take advantage of escape opportunities in the early stages of capture or when transiting to a permanent camp.[43] When the full range of available testimony is considered, it is clear that at least 25 per cent of Australian airmen held in Stalag Luft III engaged in escape or escape-related work. While more, as indicated earlier, acted disruptively, it falls short of the near-100 per cent claim.

The question arises, particularly in the light of the reprisals following the March 1944 mass breakout: how effective was Stalag Luft III's programme of active disruption? MI9's historians determined that Stalag Luft III was an escape-minded camp.[44] In retrospect, Stalag Luft III's former camp commandant acknowledged that the British—including Australians—'were the most defensive [prisoners] against the Germans'.[45] Indeed, Stalag Luft III's guards worked overtime trying to detect covert escape operations. It seems then that their unrelenting acts succeeded in disrupting Stalag Luft III's German staff. But what of the mass breakout and the cost of 50 lives—was that worth it? While the men regretted the tragic consequences, they believed the escape had been worthwhile because the Germans lost the materials they had stolen for the enterprise.[46] Within hours of the escape, a *Grossfahndung*—the highest level of search—was ordered.[47] The extensive manhunt stretched across the Reich and tied up considerable resources. Despite their evaluation, other evidence suggests the mass breakout failed to achieve the anticipated disruption.[48] The success of their disruptive acts is more intangible. Through their defiance, the airmen managed their captivity, reinforced their identity as on-duty servicemen, increased morale and self-esteem, relieved boredom, fostered group cohesion, and created the sense that they continued to be vital participants in the war. Their disruptive acts were important means by which they mitigated and ameliorated the strains of captivity in the disrupted world of the prison camp.[49]

Consequences

Active disruption had consequences. 345 of the Australian Stalag Luft III cohort survived the war.[50] Medical evidence for 128 survivors, including testimony held in Department of Veterans' Affairs files, statements provided in applications to the Prisoners of War Trust Fund, death certificates and coroners' reports, and family records, reveal the long-term physical and psychological legacy of their disruptive acts. One airman, for example, acquired tuberculosis from his work underground, which contributed to his early death. Another, who was beaten badly, developed lifelong migraines and leg and hip problems. One contracted a debilitating condition in his hands. Evidence suggests that 85 of these 128 men bore the mental scars of captivity. Some were morally troubled; a number suffered moral injury.[51] Some men developed claustrophobia; some suffered lifelong nightmares of being trapped in tunnels. At least one Australian airman grieved a lifetime over the death of a close friend in the post-escape reprisals, and his family believes that grief contributed to his suicide. At least 30 per cent of those men who suffered psychological damage had been involved in escape or escape-related work. Despite the toll of a lifetime of physical and psychological debility, one former Kriegie indicated a fierce pride in his participation in Stalag Luft III's programme of active disruption: 'Never during the whole of the War did I stop sticking my neck out'.[52] That disruptive attitude has been culturally acknowledged and lauded through the prism of the Great Escape.

After the war, Paul Brickhill, an Australian journalist and former Stalag Luft III Kriegie, was invited to write a book about the March 1944 mass breakout.[53] That book proved influential in how captivity in Stalag Luft III has been portrayed. The breakout, for example, was not known as the Great Escape until the publication of Brickhill's *The Great Escape* in 1951.[54] From that time, the book's title entered the lexicon as participants, bystanders, and the public all appropriated it to describe an event that still resonates. While many of the Australian cohorts were impressed with Brickhill's book, some were not pleased with John Sturges' 1963 film. They begrudge the Americans, the motorbike, a truly appalling Australian accent, and other factual inaccuracies.[55] Despite their distaste, both the film and Brickhill's book frame captivity in Stalag Luft III as an action-packed success story. They reinforce and validate the Kriegies' active disruption. Indeed, despite the physical, psychological, and human costs, they revelled in their defiant acts: 'We were the "naughty boys"', one former Kriegie proudly declared four decades later, 'and never stopped fighting the Germans in every way we could'.[56]

Notes

1 The author is grateful for the support of the Australian Government Research Training Program Scholarship. She also acknowledges with gratitude the Department of Veterans' Affairs (DVA) for granting special access to Second World War medical files under Section 56(2) of the Archives Act 1983 and giving permission to quote from material provided under that access. As stipulated, identifying details from DVA sources have been redacted. Part of this chapter has been adapted from an article posted on From Balloons to Drones (see '"For you the war is not over": Active Disruption in the Barbed Wire Battleground,' From Balloons to Drones, 18 December 2017, https://balloonstodrones.com/tag/kristen-alexander/) and has been reproduced with permission. Some content has also been drawn from Kristen Margaret Alexander, "Emotions of Captivity: Australian Airmen Prisoners of Stalag Luft III and their Families" (PhD Thesis, UNSW Canberra, 2020), 141–76.
2 Technically, 'airman', and its plural, 'airmen', is an air force rank. This chapter however, uses 'airman' and 'airmen' as generic terms to collectively refer to all Royal Australian Air Force (RAAF) and Royal Air Force (RAF) personnel.
3 Letter, James Catanach to William Alan Catanach, 28 March 1943, Shrine of Remembrance, James Catanach Collection, 2013.CAT050.
4 Alexander, "Emotions of Captivity," 599–601.
5 Calton Younger, *No Flight from the Cage: The Compelling Memoir of a Bomber Command Prisoner of War during the Second World War* (n.p.: Fighting High, 2013), 40; interview, Rex Austin, 5 June 2003, no. 0382, Australians at War Film Archive (AAWFA); Alec Arnel, interview by author, 29 October 2015; Bruce Lumsden, 24 June 1986, "The Complete Tour: Letters of Jaime Bradbeer and Bruce Lumsden, April 1985–October 1990" (unpublished manuscript), privately held. Note: this latter source, cited throughout, is a collection of letters written by Lumsden to Bradbeer. It is, in effect, an epistolary memoir.
6 Letter, "Usual Taunt," *Bruce Lumsden*, 24 June 1986, 'The Complete Tour'. Examples of those who heard the phrase or variations of it: wartime logbook, Ronald Baines, 8, privately held; J.R.G. (Jack) Morschel, *A Lancaster's Participation in Normandy Invasion 1944* (self-published, 1999), 32; Younger, *No Flight from the Cage*, 37; interview, Kenneth Gaulton, 3 February 2004, no. 1276, AAWFA.
7 Diary, Guy Grey-Smith, 26 January 1942, PR05675, Australia War Memorial (AWM), Canberra; wartime logbook, Robert Mills, 31, privately held; letter, Geoffrey Breadon to his parents, "Tasmanian Writes from Nazi Camp Stalagluft 3," *Mercury* (Hobart), 23 May 1944, 6; letter, Basil Kerwin to his sister, *Australian Women's Weekly*, 15 April 1944, 10; Morschel, *A Lancaster's Participation in Normandy Invasion 1944*, 47.
8 Alexander, "Emotions of Captivity," 115–27.
9 Bruce Lumsden to Jaime Bradbeer, letter 8 April 1988, 'The Complete Tour'.
10 Bruce Lumsden to Jaime Bradbeer, letter 8 April 1988 'The Complete Tour' [Original emphasis].
11 Bruce Lumsden to Jaime Bradbeer, letter 8 April 1988 'The Complete Tour'.

12 H. P. Clark, *Wirebound World: Stalag Luft III* (London: Alfred H. Cooper & Sons Ltd, 1946), 6; Peter Elliott, "Maps (and more) for the Chaps—Escape Aids and Training," *Royal Air Force Historical Society Journal* 56 (2013): 37; Bruce Lumsden to Jaime Bradbeer, letter 21 August 1989, 'The Complete Tour'; Geoffrey Cornish, interview, 2 July 2004, no. 1388, AAWFA.

13 Wartime logbook, Alex Kerr, 9, privately held.

14 Richard Winn, interview, 4 March 2004, no. 1508, AAWFA; Justin O'Byrne, interview by John Meredith, 31 October 1986, John Meredith Folklore Collection, 2991668, National Library of Australia (NLA), Canberra; Calton Younger, interview by Debbie Frith, November 2002, no. 23329, Imperial War Museum Sound Archive (IWMSA), London; radio segment, Justin O'Byrne, "*Mercury* Radio Roundsman," *Radio 7LA* (Launceston), aired c. July 1947; Horace 'Bill' Fordyce, interview, 19 June 2003, no. 0523, AAWFA; Cornish, interview; Paul Royle, interview by John Bannister, 2 December 2012, no. 26605, IWMSA.

15 Charles R. Lark, *A Lark on the Wing: Memoirs World War II and 460 Squadron,* F 940.548194 L324l, AWM, 75.

16 Paul Brickhill and Conrad Norton, *Escape to Danger* (London: Faber and Faber, 1954), 234; Gaulton, interview.

17 Fordyce, interview.

18 Letter, redacted to Deputy Commissioner, 1 June 1984, MX035229–02, Department of Veterans' Affairs (DVA), Canberra; wartime logbook, Kenneth Todd, 29, privately held; Alec Arnel, interview by author, 27 November 2014; Alan Righetti, interview, 16 September 2003, no. 0984, AAWFA; Graham Berry, interview by author, 7 June 2016; Bruce Lumsden to Jaime Bradbeer, letter 10 June 1988, 'The Complete Tour'.

19 Ron Mackenzie, *An Ordinary War 1940–1945* (Wangaratta: Shoestring Press, 1995), 56.

20 Bruce Lumsden to Jaime Bradbeer, letter 10 June 1988, 'The Complete Tour'.

21 Alec Arnel, interview by author, 9 October 2014; Berry, interview; Younger, interview; prisoner of war identification card, Rudolph Leu, A13950, LEU R, National Archives of Australia, Canberra; Justin O'Byrne, interview; Fordyce, interview.

22 Personal statement, author redacted, 17 January 1980, NCX065122–02, DVA.

23 John Herington, *Australia in the War of 1939–1945: Series Three, Vol. IV: Air Power Over Europe, 1944–1945* (Canberra: Australian War Memorial, 1963), 485.

24 Air Ministry, *The King's Regulations and Air Council Instructions for the Royal Air Force with Appendices and Index, 1942* (London: His Majesty's Stationery Office, 1942); *Air Publication 1548: Instructions and Guide to All Officers and Airmen of the Royal Air Force regarding Precautions to be Taken in the Event of Falling into the Hands of an Enemy* ([no publication details], June 1941).

25 Louise Williams, *A True Story of the Great Escape* (Crows Nest: Allen & Unwin, 2015), 205–6.

26 Drawing, Albert Henry Comber, "Flight Lieutenants (Mac) Jones and (Rusty) Kierath, RAAF at Work, Stalag Luft III, Germany, 1945," ART34781.019, AWM; British War Office, *Stalag Luft III: An Official History of the 'Great Escape' POW Camp* (Barnsley: Frontline Books, 2016), 30, 69.

27 Beecroft Probus talk, Torres Ferres, "A POW in Germany," 3 February 1989, PR90/035, AWM; Mike Netherway, interview by author, 28 September 2016; Righetti, interview; Alec Arnel, interview by author, 29 January 2015; "Gloucester Boy's Part in Great Escape," *Gloucester Advocate,* 29 May 1951, 1; Mark Whittaker, "The Men Marked X," *The Australian Magazine,* 26–27 March, 1994, 39.

28 British War Office, *Stalag Luft III,* 30, 165, 234; Kristen Alexander, "Australian Compass Makers . . . and More," *Australians in Stalag Luft III* (blog), 18 August 2016, http://australiansinsliii.blogspot.com.au/2016/08/australian-compass-makers.html; Mark Hillier, *Stalag Luft I: An Official Account of the PoW Camp for Air Force*

Personnel 1940–1945 (Barnsley: Frontline Books, 2018), 87, 94; MI9 report, Ronald Damman, 14 July 1945, AWM54, 81/4/135, AWM.

29 British War Office, *Stalag Luft III*, 161, 28, 169, 162.

30 Winn, interview; Berry, interview; Cornish, interview; George Archer, interview, "Grandstand," *ABC Radio*, April 1993.

31 British War Office, *Stalag Luft III*, 46.

32 Roberts Dunstan, "Diggers 'Relive' Years in German Prison Camp," *Herald* (Melbourne), 3 October 1951; Whittaker, "The Men Marked X," *The Australian Magazine*, 26–27 March 1994, 40; Richard Winn, *A Fighter Pilot's Diary of World War 2*. (self-pub., 2003), 52; Greg McLeod, interview by author, 15 January 2016; Winn, interview.

33 Justin O'Byrne, "Sound Recording, Reminiscential Conversations between the Hon. Justin O'Byrne and the Hon. Clyde Cameron," 29 August 1983–28 July 1984, Oral TRC 1690, Tape 51, NLA; Anne O'Byrne, interview by author, 26 May 2016.

34 Dunstan, "Diggers 'Relive' Years in German Prison Camp".

35 Archer, interview.

36 "Camp History of Stalag Luft III (Sagan) Air Force Personnel, April 1942–January 1945: Part III North Compound," WO 208/3283, The National Archives (UK), London, 49–50.

37 The duty was rescinded in the April 1944 revised and retitled *Air Publication 1548: The Responsibilities of a Prisoner of War* (3rd edition, April 1944).

38 Bruce Lumsden to Jaime Bradbeer, letter 8 April 1988, 'The Complete Tour.

39 Alexander, "Emotions of Captivity," 158–63.

40 AWM, AWM 54, 779/3/129 Parts 1–30: [Prisoners of War and Internees—Examinations and Interrogations:] Statements by repatriated or released Prisoners of War (RAAF) taken at No 11 PDRC, Brighton, letter England, 1945.

41 Justin O'Byrne also included details of his escape activity in a number of camps, including involvement in the Great Escape and Wooden Horse effort. See O'Byrne, "Sound Recording".

42 Neville Wylie, *Barbed Wire Diplomacy: Britain, Germany, and the Politics of War, 1939–1945* (Oxford: Oxford University Press, 2010), 232; Alec Arnel, interview, 27 November 2014; Bruce Lumsden to Jaime Bradbeer, letter 7 December 1987, 'The Complete Tour'; Geoffrey Bernard Coombes, interview, 23 March 1989, S00551, Keith Murdoch Sound Archive of Australia in the War of 1939–45, AWM.

43 Derrick Rolland, comps., *Airmen I have Met: Their Stories*. (Self-published, Bright: 1999), 173; Bruce Lumsden to Jaime Bradbeer, letter 5 March 1987, 'The Complete Tour'; A.L. Cochrane, "Notes on the Psychology of Prisoners of War," *The British Medical Journal* 1, no. 4442, (23 February 1946), 282; Walter A. Lunden, "Captivity Psychoses Among Prisoners of War," *Journal of Criminal Law and Criminology* 39, no. 6 (1949): 725–26.

44 M. R. D. Foot and J. M. Langley, *MI 9: The British Secret Service that Fostered Escape and Evasion 1939–1945 and its American Counterpart* (London: Book Club Associates, 1979), 108.

45 Marilyn Jeffers Walton, and Michael C. Eberhardt, *From Commandant to Captive: The Memoirs of Stalag Luft III Commandant Col. Friedrich Wilhelm von Lindeiner gennant von Wildau with Postwar Interviews, Letters and Testimony* (n.p.: Lulu Publishing Services, 2015), 63.

46 David Edlington, "The Great Escape Recalled: 60 Years on, Survivors Tell of Famous Breakout," *Air Force News: The Official Newspaper of the Royal Australian Air Force* 46, no. 4 (25 March 2004): 15.

47 "The Stalag Luft 3 War Crimes Trial Commencing at Hamburg 1 July 1947 Part 1: Voluntary Statement Max Wielen," n.d., AWM, AWM54, 1010/6/132; Guy Walters, *The Real Great Escape* (London: Bantam Press, 2013), 155–56.

48 Guy Walters, "Five Myths of the Great Escape," *History Extra*, 22 March 2019, www.historyextra.com; Walters, *The Real Great Escape*, 301–3.

49 Alexander, "Emotions of Captivity," 30.

50 Five men were shot in the 'Great Escape' reprisals. One man was killed in April 1945 by Allied strafing during the forced march.

51 For details of psychological disturbance during and after captivity, refer Alexander, 'Emotions of Captivity' chapters seven and ten. Chapter eleven considers the moral dimension of captivity.

52 Personal statement, author redacted, 17 January 1980, NCX065122–02, DVA.

53 Paul Brickhill, *The Great Escape* (London: Faber and Faber, 1951). A section on the mass escape was included in the earlier publication, Brickhill and Norton, *Escape to Danger*.

54 Stephen Dando-Collins, *The Hero Maker: A Biography of Paul Brickhill* (North Sydney: Penguin Random House Australia, 2016), 198; Brickhill, *The Great Escape*.

55 Cath McNamara, interview by author, 18 July 2016; Justin O'Byrne, sound recording; Righetti, interview.

56 Personal statement, author redacted, 17 January 1980, NCX065122–02, DVA.

References

Archival sources

Australian War Memorial

ART34781.019: Flight Lieutenants (Mac) Jones and (Rusty) Kierath, RAAF at work, Stalag Luft III, Germany.

AWM54: Written records 1939–45 War.

F940.548194 L3241: Charles R. Lark, *A Lark on the Wing: Memoirs World War II and 460 Squadron.*

PR05675: Grey-Smith, Guy (Flight Lieutenant, b.1916 – d.1981).

PR90/035: Ferres, Torres (Air Navigator), 156 Squadron (Pathfinders).

Keith Murdoch Sound Archive of Australia in the War of 1939–45, Australian War Memorial

S00551: Interview with Geoffrey Bernard Coombes.

Department of Veterans' Affairs, Canberra

MX035229–02.
NCX065122–02.

Imperial War Museum Sound Archive

Interview No. 23329: Interview with Calton Younger.
Interview No. 26605: Interview with Paul Royle.

National Archives of Australia

A13950: German Prisoner of War [POW] identification cards for captured RAAF officers, 1940–1945.

National Library of Australia

ORAL TRC 1690: Reminiscential conversations between the Hon. Justin O'Byrne and the
  Hon. Clyde Cameron.
ORAL TRC 2222/203–207: Justin O'Byrne interviewed by John Meredith for the John
  Meredith folklore collection.

Shrine of Remembrance (Melbourne)

2013.CAT050 James Catanach Collection.

The National Archives (UK)

WO 208: War Office: Directorate of Military Operations and Intelligence, and Directorate
  of Military Intelligence; Ministry of Defence, Defence Intelligence Staff: Files.

*Australians at War Film Archive, University of New South Wales–
Canberra*

Interview with Alan Righetti.
Interview with Horace 'Bill' Fordyce.
Interview with Kenneth Gaulton.
Interview with Geoffrey Cornish.
Interview with Rex Austin.
Interview with Richard Winn.

Private collections

Alex Kerr.
Kenneth Todd.
Robert Mills.
Ronald Baines.

Government sources

Air Ministry. *Air Publication 1548: Instructions and Guide to All Officers and Airmen
  of the Royal Force regarding Precautions to be Taken in the Event of Falling into the
  Hands of an Enemy.* 2nd ed. London: HMSO, June 1941.
———. *Air Publication 1548: The Responsibilities of a Prisoner of War.* 3rd ed.,
  April 1944.
———. *The King's Regulations and Air Council Instructions for the Royal Air Force with
  Appendices and Index, 1942.* London: HMSO, 1942.

Published sources

Brickhill, Paul. *The Great Escape.* London: Faber and Faber, 1951.
Brickhill, Paul, and Conrad Norton. *Escape to Danger.* London: Faber and Faber, 1954.
British War Office. *Stalag Luft III: An Official History of the 'Great Escape' POW Camp.*
  Barnsley: Frontline Books, 2016.

Clark, H. P. *Wirebound World: Stalag Luft III*. London: Alfred H. Cooper & Sons Ltd, 1946.

Cochrane, A. L. "Notes on the Psychology of Prisoners of War." *The British Medical Journal* 1, no. 4442 (23 February 1946): 282–86.

Dando-Collins, Stephen. *The Hero Maker: A Biography of Paul Brickhill*. North Sydney: Penguin/Random House Australia, 2016.

Edlington, David. "The Great Escape Recalled: 60 Years on, Survivors Tell of Famous Breakout." *Air Force News: The Official Newspaper of the Royal Australian Air Force* 46, no. 4 (25 March 2004): 15.

Elliott, Peter. "Maps (and more) for the Chaps – Escape Aids and Training." *Royal Air Force Historical Society Journal* 56 (2013).

Foot, M. R. D., and J. M. Langley. *MI9: The British Secret Service that Fostered Escape and Evasion 1939–1945 and its American Counterpart*. London: Book Club Associates, 1979.

Herington, John. *Australia in the War of 1939–1945: Series Three, Vol. IV: Air Power Over Europe, 1944–1945*. Canberra: Australian War Memorial, 1963.

Hillier, Mark. *Stalag Luft I: An Official Account of the PoW Camp for Air Force Person 1940–1945*. Barnsley: Frontline Books, 2018.

Lunden, Walter A. "Captivity Psychoses Among Prisoners of War." *Journal of Criminal Law and Criminology* 39, no. 6 (1949): 721–33.

Mackenzie, Ron. *An Ordinary War 1940–1945*. Wangaratta: Shoestring Press, 1995.

Morschel, J. R. G. *A Lancaster's Participation in Normandy Invasion 1944*. Self-published, 1999.

Rolland, Derrick, comp. *Airmen I have Met: Their Stories*. Self-published, Bright: 1999.

Walters, Guy. *The Real Great Escape*. London: Bantam Press, 2013.

Walton, Marilyn Jeffers, and Michael C. Eberhardt. *From Commandant to Captive: The Memoirs of Stalag Luft III Commandant Col. Friedrich Wilhelm von Lindeiner gennant von Wildau with Postwar Interviews, Letters and Testimony*. N.p.: Lulu Publishing Services, 2015.

Williams, Louise. *A True Story of the Great Escape*. Crows Nest, NSW: Allen & Unwin, 2015.

Winn, Richard. *A Fighter Pilot's Diary of World War 2*. Self-published, 2003.

Wylie, Neville. *Barbed Wire Diplomacy: Britain, Germany, and the Politics of War, 1939–1945*. Oxford: Oxford University Press, 2010.

Younger, Calton. *No Flight from the Cage: The Compelling Memoir of a Bomber Command Prisoner of War during the Second World War*. N.p.: Flying High, 2013.

Theses and unpublished manuscripts

Alexander, Kristen. "Emotions of Captivity: Australian Airmen Prisoners of Stalag Luft III and their Families." PhD thesis, UNSW Canberra, 2020.

"The Complete Tour: Letters of Jamie Bradbeer and Bruce Lumsden, April 1985–October 1990." Unpublished Manuscript, Privately Held.

3 Sir James Rowland and the changing strategic use of air power in Australia, 1942–1979

Peter Yule and Nicole Townsend

Many celebrated chiefs have led the Royal Australian Air Force (RAAF) since its formation in 1921, but since the 1950s, their careers have generally followed a highly predictable course. With one exception, every Chief of the Air Staff (CAS, now called the Chief of Air Force) has been a General Duties pilot, and almost all appointees have served as Air Commander or Deputy Chief of the Air Staff/ Deputy Chief of Air Force first.[1] The odd man out is Air Marshal Sir James 'Jim' Rowland, who became the first and only engineer to command the RAAF as the CAS from 1975 to 1979.[2] Rowland would also be the first CAS to command the Air Force in a legal sense after the dissolution of the Australian Air Board in 1976. Until then, the Air Board had been the controlling body of the Air Force, while the CAS functioned as the operational head of the Air Force and chairman of the Air Board. As a bomber pilot and pathfinder, prisoner of war, aeronautical engineer, test pilot, Air Member for Technical Services, CAS, and finally Governor of New South Wales, Rowland had a distinguished and exceptional career.[3]

Rowland joined the RAAF in May 1942 and, as a Lancaster pilot in the Royal Air Force's Bomber Command, took part in the Allied attempt to defeat Nazi Germany through the strategic use of air power. In the 1930s, it became accepted dogma that 'the bomber will always get through' and that war would see the widespread destruction of cities, with the winning side being the one that wreaked the greatest destruction most rapidly.[4] However, the Battle of Britain revealed that bombers were extremely vulnerable to modern fighters, to the extent that daylight operations could only be conducted at a heavy cost. The experience of American daylight bombing raids on Germany confirmed that even heavily armed and armoured bombers could not operate effectively in hostile airspace. Short of alternative strategies to take the offensive against Germany, Britain opted to concentrate a high percentage of its resources on building a vast force of bombers designed for night operations against German cities—cities being the smallest target they could reliably hit. Enormously expensive in lives and money and highly successful in reducing German cities to ruins, the British air offensive's contribution to final victory is still a fiercely debated issue.[5]

The advance of military aviation during the Second World War was prodigious. In 1939, many air forces were still flying plywood and canvas biplanes, but by 1945, the first jets were flying over Europe, and supersonic speeds were within

DOI: 10.4324/9781003230656-5

reach. Rapid progress continued in the post-war decades, driven by the Cold War with the Soviet Union. As a test pilot and aeronautical engineer, James Rowland was at the centre of these advancements, being either the first or among the first to fly all the new aircraft types that came into service in Australia and many American and British aircraft. At this stage of his career, his responsibility was to ensure that aircraft could perform the tasks required. Australia's front-line fighter aircraft in the 1950s and 1960s, the American-built F-86 Sabre and the French Mirage, saw minimal operational service, but other aircraft such as the Lincoln bombers (the next iteration of the Second World War Lancaster) and Dakotas (designed before the war) saw operational service in Malaya. Canberra bombers, Iroquois helicopters, Hercules, and Caribou freighters operated in Vietnam. In Malaya and Vietnam, the RAAF supported army operations against Communist guerrillas in line with a forward defence strategy against a perceived Communist threat. However, there was a complete reversal of Australian defence thinking following the withdrawal of Australian troops from Vietnam and the election of the Whitlam Labor government. Forward defence was abandoned, and a new policy of 'Fortress Australia' was adopted with an emphasis on defending the northern approaches to Australia.[6]

After his unexpected elevation to the position of CAS, Rowland led the RAAF as it adapted to its greatly changed role in Australia's defence strategy. He played a crucial role in developing the concept of operations for the F-111s that were coming into service, quickly appreciating that they would give Australia a significant edge over any potential enemy for decades to come. During his tenure, Rowland also steered the RAAF through a period of great change as Australian strategic policy shifted its focus from forward defence to the 'Fortress Australia' policy, and the Department of Defence underwent significant restructuring. This chapter traces how Australia's employment of strategic air power evolved, Rowland's role in this process, and the challenges he and the RAAF faced from 1942 to 1979 as the geopolitical environment and strategic policy changed.

Fascinated by flight

James Rowland was born in 1922 and grew up on the family farm in Armidale, New South Wales. Although his family was comfortable enough financially, it was a spartan upbringing, without electricity and with a high level of self-sufficiency. When he was eight or nine, he saw his first aeroplane, flying over the farm—'a tiny biplane putt-putting its way south'. Rowland later recalled that he quickly lost interest in ships after that; he was 'fascinated by flight', which remained the 'last frontier to be explored' by man.[7] His ambition became to design aircraft – 'sleek and beautiful things with flowing lines' that could reach great speeds and carry far more passengers and cargo over much greater distances than the aircraft of the 1930s.

While 14 years old, Rowland sent his design for a rear-firing machine gun for the Hawker Fury to the Hawker company. He received a considered reply, pointing to the possible impact of the suggestion on the aircraft's balance. A bright teenager, Rowland attended Sydney's prestigious Cranbrook School on a scholarship before enrolling to study aeronautical engineering at the University of Sydney in 1940. Rowland continued to be a good student, but his mind seems to have been focused more on the critical air battles on the other side of the world than on his studies; his lecture notes were adorned with sketches of Spitfires, Hurricanes, and Me 109s (Figure 3.1).[8]

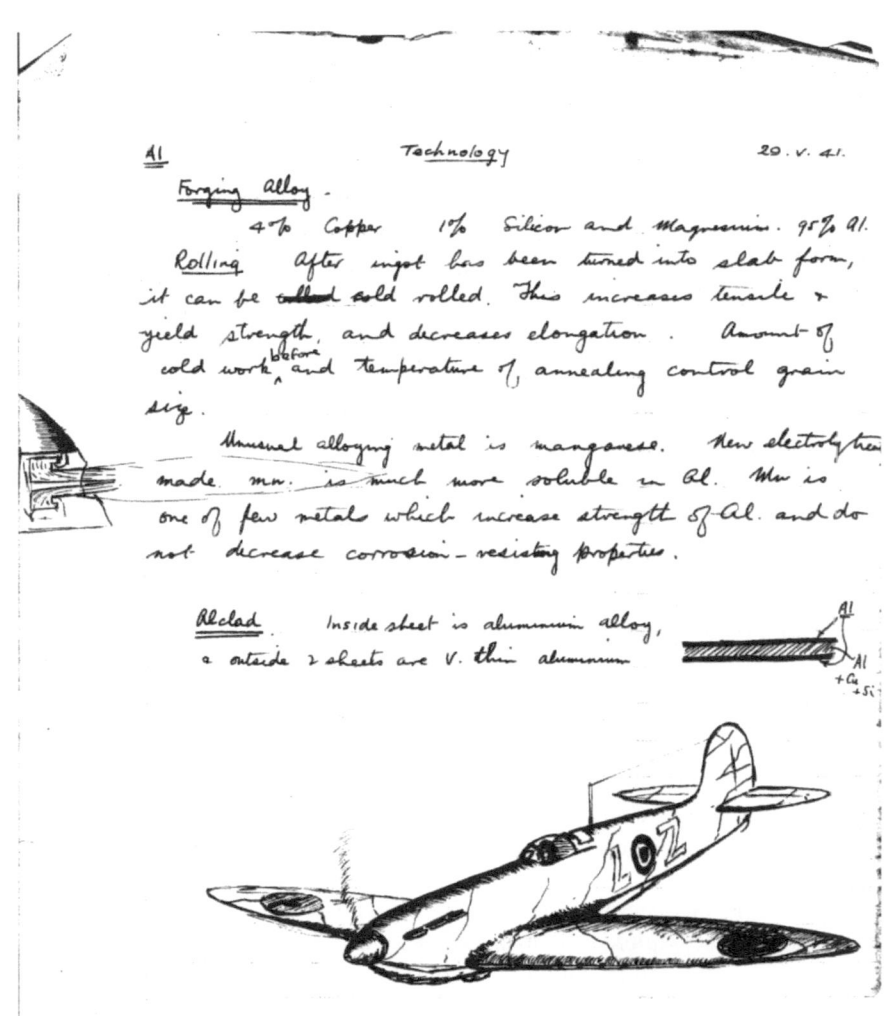

Figure 3.1 A page from James Rowland's lecture notes, illustrated with a hand-drawn sketch of an Me 109. Courtesy of Anni Rowland-Campbell.

From the outbreak of the Second World War, Rowland had been determined to join the RAAF and become a fighter pilot, but his parents refused to allow him to join until he was nineteen. By then, the manpower authorities had ruled that engineering students could not enlist until they finished their courses. To get around this ruling, Rowland and some of his friends deliberately failed their second-year exams so they would be expelled from the course, allowing them to enlist. When Rowland joined the RAAF in late 1941, the need for fighter pilots was lessening, and the Empire Air Training Scheme was training aircrew for the bombing offensive against Germany. Accordingly, he was sent first to the No. 10 Elementary Flying School at Temora, New South Wales, to fly Tiger Moths, then to Canada for further training on Harvards and finally to England, where he completed his training on Oxfords, Wellingtons, and, eventually, four-engine Halifaxes and Lancasters. Rowland chose to join the RAF rather than a RAAF squadron, and his crew included young men from around Britain and the Commonwealth. Rowland was quickly recognised as an outstanding pilot, and before he flew on operations, he and his crew were selected for the elite Pathfinder force. This required committing to a tour of forty-five operations rather than the standard 30, but as few crews survived the 30 operations tour, this appeared to make little difference.

Rowland became a small cog in the mighty machine of the bombing offensive against Germany. The rationale, as well as the morality of the bombing offensive, has often been queried. Still, in 1940, when Britain was isolated and without an ally (France capitulated in May 1940, and the Soviet Union and the United States had not yet entered the war), it appeared to be the only way to attack Germany directly. Consequently, the British Government decided in August 1940 to allocate a large proportion of its limited resources to building bombers and training bomber crews.[9] Initially, British bombing focused on strictly military targets, but an analysis of bombing results in 1941 showed that bombs hardly ever hit their targets—in fact, on many occasions, the Germans could not even guess what target the British bombers were trying to hit.[10] The smallest target they could hit fairly reliably was a city, so the decision was made to flatten German cities, which they believed would 'de-house' factory workers, cripple industry and transport, and destroy German morale. The effectiveness of this campaign is still debated, but at the very least, it led to an enormous diversion of German resources into air defence. Whether that outweighed the resources the British devoted to building bombers and training crews is unclear. However, one consequence of the bombing campaign was crystal clear—approximately 55,000 aircrew of Bomber Command were killed, including 3486 Australians.

Rowland and his crew flew 34 operations between August 1944 and January 1945, and he was awarded the Distinguished Flying Cross (DFC) for completing an operation after an engine failure. However, on 6 January 1945, during a raid on Hanau near Frankfurt, his aircraft was accidentally rammed by a Canadian-crewed Halifax on its first operation. Rowland was the only survivor from the two aircraft, although recent research shows several of his crew parachuted safely to the ground, only to be murdered by German police.[11] Rowland was fortunate to survive capture by the Gestapo but spent the rest of the war as a prisoner of war in Stalag XIII-D, near Nuremberg.

After repatriation in August 1945, Rowland returned to university to finish his aeronautical engineering course and then, in 1947, joined the air force again. His flying skills, operational experience and engineering knowledge stamped him as a likely test pilot, so he was sent to the Aircraft Research and Development Unit (ARDU), based at Laverton, Victoria. In a move that would have surprising ramifications for Rowland many years later, most ARDU test pilots were transferred from the General Duties branch to the Technical Branch in September 1948. Rowland's training and promotions were then achieved in the Technical Branch (later renamed the Engineer Branch). In 1949, he became only the second Australian to train at the Empire Test Pilots School in England. The school's professed objective was to produce 'the most highly qualified test pilots in the world'. Graduates would be 'superb' flyers and capable of analysing the performance and construction of all aircraft they handled, which was no small task, given the remarkable variety of aircraft Rowland and other pilots flew during their time at the school.[12] Rowland was the first Australian to fly a Canberra bomber and was also one of the first to experience the transonic zone when flying the Supermarine 510, a forerunner of the Supermarine Swift.[13]

For most of the 1950s, Rowland remained at ARDU, rising to lead its Research and Development Squadron. In this period, ARDU was at the forefront of research to resolve the many new problems that appeared with the rapid progression of aviation technology since 1939. Some of ARDU's work in the early 1950s was also at the forefront of international flight research and development, including its boundary layer separation trials on Mustangs and Vampires. The unit also led the way in dealing with the challenges of introducing the F-86 Sabres and the Canberras into service. The Australian versions differed significantly from their American and British originals, and they also operated in significantly different conditions, presenting the RAAF with unexpected challenges. Both aircraft, for example, experienced engine stalls at high altitudes in the unusually cold atmosphere over Darwin, a unique meteorological phenomenon that Rowland was among the first to observe. Rowland was also responsible for designing the fix to the unfortunate tendency of Sabres to decapitate ejecting pilots—a problem the Americans knew about but did not think worth fixing.[14]

Rowland was involved from an early stage in the selection of a fighter to replace the Sabres. He was one of the strongest opponents of buying the F-104 Starfighter, which was in retrospect a wise decision, given the disastrous safety record of that aircraft in service with several European air forces, as well as its 'considerable limitations in all but its air intercept role'.[15] In November 1960, the government decided to buy the French Mirage and Rowland was sent to Paris for three years as the senior engineer on the procurement team. The Mirage was the first aircraft the RAAF did not buy from Britain or the United States. It was also chosen very early in its development, which presented many challenges for the procurement team. Rowland handled the complex technical issues with great skill, and his fluent French saved the project from several disasters.[16]

The path to leadership

By the early 1960s, Rowland was picked out by his superiors as having all the qualities needed to rise to the top of the Engineer Branch. After returning to Australia from France, he had a succession of postings designed to give him wide command experience and expose him to varying challenges. These included postings to the Department of Air in Canberra as AIR ENG 1, No. 3 Aircraft Depot at Amberley, Queensland, as the commanding officer, and RAAF Operational Command at Glenbrook, New South Wales, as the senior engineering staff officer.

After the successful introduction of the Mirage, the RAAF's major engineering challenge for the next decade was the introduction of the F-111 and Rowland was heavily involved from the outset in dealing with the numerous teething problems of that brilliant aircraft. Until this time, virtually all Rowland's experience and training had been in test flying and engineering. Issues of strategy and doctrine had been peripheral, but in 1971, he spent a year in London at the Imperial Defence College. There, he studied an intensive course on international security and 'strategic vision', which was ideal for lifting his focus from the details of aircraft engineering to wider questions of strategy and the use of military power.

When he returned to Australia, Rowland was appointed SMAINTSO (the eloquent military acronym for the senior maintenance staff officer) at Support Command Headquarters in Melbourne. However, he had barely started in that role before being promoted to Director General of Aircraft Engineering in Canberra. Then came what was expected to be his last promotion, to Air Member for Technical Services in November 1972. There was no reason to think that he would rise further. This position was the highest in the Engineer Branch, and there had never been a Chief of the Air Staff who was not a General Duties pilot; in fact, the regulations insisted on it.[17] There was no reason to expect that this would change when Charles Read retired in 1975, therefore Rowland and his colleagues expected that he would head the Engineer Branch until his retirement. However, late in 1974, Lance Barnard, the Minister for Defence in the Whitlam Government, nominated Rowland as the next CAS.

The reasons for the choice were never clearly stated. However, it is reasonably certain that the decision was made by Sir Arthur Tange, the Secretary of the Department of Defence, as a part of his crusade to remake the department. Tange believed that Rowland would have the flexibility of thought and the personal qualities to lead the RAAF through the inevitable turmoil that would follow the restructuring of the defence bureaucracy. Many RAAF officers rejected the reforms out of hand, and Tange likely deduced that Rowland would find ways to work within the new system rather than blindly fighting against it.[18] Tange was also aware that the RAAF had some 'perplexing problems' that stemmed from its highly advanced technology and the vulnerability of air safety and combat readiness to any shortcomings related to engineering maintenance and supply.[19] Tange probably believed that an engineer would be better able to deal with these problems than the General Duties officers who possessed little experience in these areas.

Unsurprisingly, Rowland's appointment met with significant opposition. His appointment led to a revolt by the General Duties officers on the Air Board, who threatened to block Rowland's transfer from the Engineer Branch to the General Duties Branch, leading to perhaps the most extraordinary meeting in the history of the Air Board. At the meeting, the Secretary of the Department of Air, Frederick Green, gave the board members (less Rowland) a stern lecture about their duty and political reality. The minutes of the Air Board are generally bland records of issues discussed and decisions agreed. However, the board's 20 November 1974 meeting minutes made it clear that harsh words were spoken and strong feelings expressed—but the upshot was that the Air Board rolled over, and Jim Rowland became CAS.[20]

The job was never easy, yet Rowland faced a combination of challenges which no previous peacetime CAS had encountered. The mid-1970s saw a major reorientation of Australia's strategic posture, a complete restructure of the Department of Defence that involved radical changes in how the RAAF was run, and Australia's most serious economic downturn since the 1930s, which led to high unemployment, rampant inflation, and constant pressure on the defence budget. Rowland had to lead the RAAF in developing new doctrines, structures, and policies appropriate for the new strategic environment, all while managing within a continually declining budget (in real terms) and negotiating his way through the stifling committee structure of the new system. Although Rowland made no secret that he would rather be flying than sitting in meetings, the perspective of time shows that he rose to these challenges with remarkable success.

Fortress Australia

Soon after Rowland became CAS, the RAAF was called on to evacuate refugees from South Vietnam as the North Vietnamese army marched toward Saigon. The fall of South Vietnam marked the symbolic end of Australia's post–Second World War policy of 'forward defence' based on acting with Britain or the United States to confront possible threats far from Australia. This policy had driven Australia's commitment of forces to Korea, Malaysia, Borneo, and Vietnam; however, the British withdrawal from east of Suez and the disaster of Vietnam led to a major shift in Australian policy. As with the reform of the Department of Defence's structure, much of the impetus for rethinking Australian defence policy came from Sir Arthur Tange, with the support of the Whitlam Government.

The new policy of 'Fortress Australia' emphasised the defence of mainland Australia, with a focus on commanding the air–sea gap between Australia and Southeast Asia. Inevitably, this policy led to priority being given to the Navy and Air Force, with the Army's role reduced to 'mopping up' small groups of enemy forces that might succeed in crossing the air–sea gap.[21] It also led to a reorientation of Australia's armed forces from the country's southeast to the north and west, where they could defend the air–sea gap. Previously, the armed forces had been focused on Australia's southeast, which was convenient for deployments abroad. The new policy combined with the assessment that Australia faced no foreseeable threat for the next ten to fifteen years—quite sensible as it turned out, with both China and Indonesia becoming less bellicose—but this became a real concern for Rowland and the other defence chiefs, primarily because it justified steep budget cuts.

Jim Rowland became CAS unexpectedly and had previously been immersed in practical matters of aeronautical engineering rather than policy formulation. Still, he quickly appreciated the new paradigm facing the RAAF and began the process of adaptation at both the theoretical and practical levels. He realised that it was essential to develop new operational concepts suited to the new strategic environment and appointed David Evans as Chief of Air Force Operations to develop the RAAF's doctrine for the new era. Although the final version was not approved until shortly after Rowland's retirement, many concepts were already in place. A good example of this was Rowland's early advocacy for developing RAAF Base Tindal as the RAAF's main northern base. Located more than 320 kilometres south-east of Darwin, it would be less vulnerable to enemy attack and cyclones than the Northern Territory capital.[22]

Even more to the point was the concept of operations developed for the air force's new F-111s. The problems experienced by the early F-111s in America led Rowland's predecessor, Charles Read, to demand that Australia's F-111s be handled with extreme care, leaving pilots unable to explore the aircraft's potential. One of Rowland's first actions after taking over from Read was to lift these restrictions. Rowland saw the F-111's potential, understood its technical problems, worked to fix them, and led the RAAF in learning how to use these new aircraft. The original decision to buy the F-111 was made when a bellicose Indonesia appeared to pose a serious threat to Australia, and the F-111 was seen primarily as a deterrent to Indonesian aggression. The fall of President Sukarno in 1965 led to a complete transformation in Australian relations with Indonesia. By the 1970s, relations were friendly. It became clear that Indonesia had neither the intent nor the ability to threaten Australia. So, what then was the purpose of the F-111s and how were they to be used?

Rowland began a new debate on how and where the F-111s should be employed. In 1976, he outlined his views on the role of the F-111: this aircraft was the 'main element' in Australia's offensive air capability, and without them, Australia would be incapable of retaliating against or deterring any potential aggressor. Without a developed and maintained offensive capability, Australia would not be able to deter potential enemies from acting against the Australian national interest.[23] Almost without realising it, in the F-111, the RAAF found itself in possession of a formidable offensive weapon for the first time since the Second World War, and Rowland encouraged the operators to test their capabilities. He allowed the F-111s to be employed to the full extent of their capabilities in exercises in Australia and overseas to give pilots valuable training in operating against modern fighters and surface-to-air missiles and within 'unique' electronic warfare environments. Differing climates and terrains also enabled refinement of tactics, techniques, and procedures.[24] By the time Rowland retired in 1979, the RAAF was operating the F-111s capably and confidently; they would give Australia a military edge in Southeast Asia for many years to come.

Deteriorating and missing capabilities

Despite their obvious capabilities, the F-111Cs had serious deficiencies in weapons and reconnaissance capability when they arrived in Australia. When they arrived, they could only carry Mk-80 series unguided bombs, which were only slightly different technologically from those Rowland's Lancaster dropped on German targets 30 years earlier. Rowland continually pressed for the F-111s' weapons to be upgraded, but the defence bureaucracy slowed progress, and it was not until the 1980s that the aircraft were fitted with precision-guided weapons.[25]

The strategic concentration on Australia's northern air–sea gap required the ability to see whatever might be attempting to cross the gap. The F-111Cs did not initially have a reconnaissance capability, although this capability had been specified in the original Air Staff requirement in the early 1960s; its absence was due to the US Air Force changing its requirements, which led to the proposed specialised reconnaissance F-111 being cancelled. Rowland was determined that the RAAF should develop its reconnaissance capability for the F-111s, so four F-111s were fitted with a suite of optical and infrared cameras and related equipment. The first was completed by April 1979, and the remaining three during 1980. The project came in on time and within budget.[26]

The RAAF leadership in the 1970s has been criticised for having an 'equipment replacement mentality', implying that little or no thought went into the requests for new aircraft and other equipment. This criticism ignores the fact that no project could get through the bureaucratic approval system without exhaustive examination of every aspect, including the strategic justification. Additionally, there is no doubt that many of the aircraft the RAAF acquired in the 1950s and 1960s had become obsolescent, and a clear strategic case could be made for their replacement. More than a decade after his retirement, Rowland outlined the situation the RAAF then faced in his 1992 talk to a RAAF history conference:

> In 1975, many of our aircraft were getting a bit long in the tooth. The Mirage was doing well, but fatigue was starting to appear. In fact, our aircraft had become the lead aircraft in terms of hours, and they had worked hard and given us a wonderful run. In 1976, I flew the new F-15 from Scott Air Force Base in the United States, and it was clear that a quantum jump in capability had been made. In any case, the old Mirage's armament was limited. We got Magic to improve it in air to air, but in air-to-ground it could not really offer much, and it was only a matter of time before something would have to be done. The F-111 was a very good aircraft, but it had only iron bombs, and its analog weapon system was limited with regard to new weapons, as the USAF had gone digital. The [Hercules] C-130As ran out of fatigue life, and were about to be grounded. Corrosion problems began to beset the [Orion]

P3B's, and their anti-submarine gear had become dated. We had had the Caribous and the Iroquois helicopters since about 1956, and the early models were beginning to give some concern. So there was more to be done than the budget would provide. My four years as CAS produced the C-130H and the P3C, and yet another fighter mission. The Mirage replacement had to wait. Other clear needs were a flight refuelling capacity, and AEWC [Airborne early warning and control][27]

Conclusion

Although Rowland had not followed an orthodox path to become CAS and his appointment had been resisted by many of his peers, his outgoing personality, strong people skills, and a level of professional expertise that went well beyond his engineering training enabled him to guide the RAAF successfully through an extremely turbulent period in its history. Before his retirement in 1979, the RAAF had begun looking for a replacement for the ageing Mirages, and Rowland flew some of the candidates, including the F-15 Eagle. Later, as Governor of New South Wales, he called in some favours to fly an F/A-18 Hornet. Rowland's career in aviation began with a primitive Gypsy Moth biplane in 1940. Over the next half-century, he flew almost every aircraft type in service in the RAAF and many other British and American aircraft. Few people would have been so familiar with the changing capabilities of military aircraft. At many different levels—bomber pilot, test pilot, aircraft engineer, and head of the RAAF—Rowland was an active participant in the changing strategic use of air power.

Notes

1 Alan Stephens, *Going Solo: The Royal Australian Air Force 1946–1971* (Canberra: Australian Government Publishing Service, 1995), 296–97.

2 Air Marshal Geoff Brown, who studied engineering as an undergraduate, served as Chief of Air Force between 2011 and 2015. However, he was never employed as an engineer.

3 The major source for this chapter is James Rowland and Peter Yule, *Pathfinder, 'Kriegie' and Gumboot Governor: The Adventurous Life of Sir James Rowland, AC, KBE, DFC, AFC* (Canberra: RAAF History and Heritage Branch, 2018). Before his death, Sir James Rowland wrote an account of his life up to 1947, together with disconnected chapters on his later life. The first author of this chapter edited Sir James' work and completed the rest of the story of his life.

4 This now famous phrase was spoken by Stanley Baldwin, who at the time of his speech had been the Prime Minister of Britain twice already but currently held the position of Lord President of the Council and leader of the Conservative Party. See House of Commons, Debates, 10 November 1932, vol. 270, https://api.parliament.uk/historic-hansard/commons/1932/nov/10/international-affairs.

5 The literature on the allied bombing campaign against Germany is vast. A classic account remains Max Hastings, *Bomber Command* (London: Michael Joseph, 1979). An excellent analysis of the origins of the campaign is Tami Davis Biddle, "British and American Approaches to Strategic Bombing: Their Origins and Implementation

in the World War II Combined Bomber Offensive," *Journal of Strategic Studies* 18, no. 1 (1995): 91–144. For arguments over the effectiveness of the strategic bombing campaign, see Sir Arthur Harris, Bomber Offensive, (London: William Collins, 1947); Kenneth P. Werrell, "The Strategic Bombing of Germany in World War II: Costs and Accomplishments," *The Journal of American History* 73, no. 3 (1986): 702–13.

6 For a detailed discussion of 'Fortress Australia' versus 'forward defence' in relation to Australian defence policy, see: Paul Dibb, "The Self-Reliant Defence of Australia: The History of an Idea," in *History as Policy: Framing the Debate on the Future of Australia's Defence Policy*, eds. Ron Huisken and Meredith Thatcher (Canberra: ANU E Press, 2007), 11–26.

7 Rowland and Yule, *Pathfinder*, 31.

8 Ibid., 51–68.

9 Michael Postan, *British War Production* (London: Longmans Green, 1952), 124, 128; Corelli Barnett, *The Audit of War: The Illusion and Reality of Britain as a Great Nation* (London: Macmillan, 1986), chap. 8.

10 The assessment was made by David Bensusan-Butt, a civil servant with the war cabinet secretariat, who analysed over 600 photographs taken during night bombing in June and July 1941. Bensusan-Butt concluded that only about five per cent of bombers setting out from Britain bombed within five miles (eight kilometres) of their target; see letter, David Bensusan-Butt to Lewin Duggan, "The Butt Report," 18 August 1941, The National Archives (TNA), AIR 14/1218. See also, Norman Longmate, *The Bombers: The RAF Offensive against Germany, 1939–1945* (London: Hutchinson, 1983), 120–25. After the war Bensusan-Butt migrated to Australia where he was an economist at the ANU for many years.

11 Rowland and Yule, *Pathfinder*, chap. 18.

12 *The Times*, 25 February 1946.

13 Ken Llewelyn, "Powdered Custard Days Are Over," *RAAF News*, 1 September 1987, 3.

14 Nev Williams, "Sabre Ejection Seat," *RAAF Radschool Association Magazine*, February 2011, 13; Minute, "Sabre Modification No 181 – Ejection through Canopy – Provisions – Publication Of," 23 June 1960, A705, 150/8/2359 PART 1, National Archives of Australia (NAA), Canberra.

15 Alan Stephens, *Going Solo: The Royal Australian Air Force 1946–1971* (Canberra: Australian Government Publishing Service, 1995), 355.

16 M.R. Susans, ed., *The RAAF Mirage Story* (Point Cook, Victoria: Royal Australian Air Force Museum, 1990).

17 The regulations as they stood in 1975 are held by the NAA. See: "Air Force Regulations 25/2 – 'Appointment of the Chief of the Air Staff'," A5034, SR1974/3068, NAA.

18 See letter, Charles Read to Arthur Tange, 14 November 1974, Sir Arthur Tange Papers, MS9847/5/7, National Library of Australia (NLA), Canberra; letter, Arthur Tange to Charles Read, 28 October 1974, Sir Arthur Tange Papers, MS 9847/5/7, NLA.

19 Sir Arthur Tange, *Defence Policy-Making: A Close-Up View, 1950–1980*, ed. Peter Edwards (Canberra: ANU E Press, 2008), 67.

20 *Minutes of the Air Board*, 20 November 1974, RAAF History and Heritage Branch.

21 See, for example, memorandum, Sir Arthur Tange, "Evolution of Defence Outlook, 1942–1988," n.d., Sir Arthur Tange Papers, MS 9847/5/7, NLA.

22 David Evans, *Down to Earth: The Autobiography of Air Marshal David Evans* (Canberra: RAAF Air Power Development Centre, 2011), 207.

23 Speech, Air Force Association Biennial Conference, August 1976, Rowland Family Archives.

24 Mark Lax, *From Controversy to Cutting Edge: A History of the F-111 in Australian Service* (Canberra: RAAF Air Power Development Centre, 2010), 133.

25 Ibid., 142–45.

26 Ibid., 138–42.

27 "Some Notes by Air Marshal Sir James Rowland," RAAF History Conference, 1992, 11.

References

Archival sources

National archives of Australia

A5034 Attorney-General's Department, Classified correspondence files, 1973–1974.
A705: Department of Defence, Correspondence files, 1912–1988.
National Library of Australia.
MS 9847: Papers of Sir Arthur Tange, 1929–2001.

The National Archives (UK)

AIR 14: Air Ministry: Bomber Command: Registered Files.

Government publications

Parliamentary Debates, Commons (UK), 5th series (1909–81).

Published sources

Barnett, Corelli. *The Audit of War: The Illusion and Reality of Britain as a Great Nation*. London: Macmillan, 1986.

Biddle, Tami Davis. "British and American Approaches to Strategic Bombing: Their Origins and Implementation in the World War II Combined Bomber Offensive." *Journal of Strategic Studies* 18, no. 1 (1995): 91–144.

Dibb, Paul. "The Self-Reliant Defence of Australia: The History of an Idea." In *History as Policy: Framing the Debate on the Future of Australia's Defence Policy*, edited by Ron Huisken and Meredith Thatcher, 11–26. Canberra: ANU E Press, 2007.

Evans, David. *Down to Earth: The Autobiography of Air Marshal David Evans*. Canberra: RAAF Air Power Development Centre, 2011.

Harris, Arthur. *Bomber Offensive*. London: William Collins, 1947.

Hastings, Max. *Bomber Command*. London: Michael Joseph, 1979.

Lax, Mark. *From Controversy to Cutting Edge: A History of the F-111 in Australian Service*. Canberra: RAAF Air Power Development Centre, 2010.

Longmate, Norman. *The Bombers: The RAF Offensive against Germany, 1939–1945*. London: Hutchinson, 1983.

Postan, Michael. *British War Production*. London: Longmans Green, 1952.

Rowland, James, and Peter Yule. *Pathfinder, 'Kriegie' and Gumboot Governor: The Adventurous Life of Sir James Rowland, AC, KBE, DFC, AFC*. Canberra: RAAF History and Heritage Branch, 2018.

Stephens, Alan. *Going Solo: The Royal Australian Air Force 1946–1971*. Canberra: Australian Government Publishing Service, 1995.

Susans, M. R., ed. *The RAAF Mirage Story*. Point Cook, Victoria: Royal Australian Air Force Museum, 1990.

Tange, Sir Arthur. *Defence Policy-Making: A Close-Up View, 1950–1980*. Edited by Peter Edwards. Canberra: ANU E Press, 2008.

Werrell, Kenneth P. "The Strategic Bombing of Germany in World War II: Costs and Accomplishments." *The Journal of American History* 73, no. 3 (1986): 702–13.

4 Australian air power strategy, technologies, and counter-insurgency in Malaya during the Cold War

Peter Hunter

This chapter considers the far-reaching consequences of the Malayan Emergency on the development of Australian air power strategy. It examines how the Royal Australian Air Force (RAAF) adapted to the new and challenging experience of counter-insurgency warfare, the methods it developed, and the pivotal relationship between military technology and air power strategy at the time. While important lessons were learnt about the need for improved technologies to overcome local geographic and environmental challenges, the far-reaching implications of the relationships between politics, ideologies, coercion, and military effects as they applied to irregular warfare, and indeed to grand strategy, were not fully appreciated.

Australian cold war strategy in Southeast Asia: forward defence

One of the most fundamental concerns for Australia throughout the Cold War was the need to balance two competing perspectives on national security. On the one hand was the desire to enlist the support of allies to help defend Australia and its maritime approaches in the face of a volatile region.[1] On the other was the need to exercise independence in foreign and defence policy, and to possess the supporting military capabilities that this would require. But such national security goals seemed contradictory, if not out of reach, given Australia's small population and limited resources. Australia could only sustain relatively small armed forces, so national defence seemed to require a close relationship with an ally who shared Australia's interests in the region and possessed the military strength necessary for protecting those interests.[2] While Australia had looked to the United Kingdom (UK) for that support during the first half of the 20th century, the experience of the Second World War had underscored the indispensability of the United States to Australian security interests.

The Australian Government's concern to draw the United States closer had indeed been a fundamental consideration in becoming involved in the Korean War. The Pacific Pact (which later became the ANZUS Treaty), negotiated by Minister for External Affairs Percy Spender off the back of the goodwill earned through Australia's early commitment to the Korean War, became a defining

DOI: 10.4324/9781003230656-6

feature of Australian security policy.[3] That said, although the Australian Government recognised the ascendancy of US military influence in its region by the early 1950s, it was not yet entirely ready to dismiss the UK's significance to Southeast Asian security. Even as the UK's commitments in Southeast Asia began to diminish as its resources and focus were diverted more to Europe, it nevertheless continued to play a significant role, particularly in its former colonial interests.[4] Therefore, Australia believed that a continuing British presence in the region was to be encouraged.[5]

Striking the right balance between the need for close alliances and the importance of self-reliance presented tough policy choices for the Australian Government. Notwithstanding the importance Australia attached to its alliances, neither the American nor the British commitment to regional security issues could be taken for granted. Thus, independence in Australian foreign and defence policy-making became more significant. Wrestling with these competing themes in the 1950s and 1960s, the Australian Government adopted the strategic concept of 'forward defence'. That strategy appeared to rest on the premise that the Australian military should be positioned to conduct operations from bases in Southeast Asia to thwart any threats to national security before they came too close. That is, to 'fight them up there before we fight them down here'.[6]

But the forward defence strategy was a more nuanced approach to dealing with the complexities of regional engagement than this rendition might suggest. Plainly, alliances and multilateralism would need to be essential elements in accounting for the complexities of Cold War politics in Southeast Asia. Forward defence stressed the need for Australian forces to work among coalitions of American, British, and other regional forces to bolster Southeast Asian security. Indeed, Prime Minister Robert Menzies advised his senior Cabinet ministers that

> the only sensible approach was to have forces that were organised and ready to move rapidly to oppose the spread of communism in Southeast Asia and which also were equipped and trained to be compatible with the emerging major Western power, the Americans.[7]

This view was underscored in the 1956 *Strategic Basis of Australian Defence Policy*, which stressed that 'Australia is not able to defend herself unaided against a major power and is dependent on the Western Powers, in particular the United States, for her ultimate security'.[8]

That said, the Australian Government remained clear-eyed about the mutability of US commitments to Australian security needs, and so the government considered it should only commit military forces to confront security risks in the immediate region. Defence officials assessed that 'the first line of Australia's defence lies in Southeast Asia, and no major threat to her security can develop, nor is she likely to be a primary objective of a major communist power, whilst Malaya is held'.[9] Malaysia's strategic significance to Australia's security was not new. It had been painfully evident during the Second World War, when, for much of 1942, Australia feared Japan was poised to invade Australia from the Malayan peninsula. The policy of forward defence, and for that matter, the establishment in 1949 of the Australia-New-Zealand-Malaya (ANZAM) defence arrangements, no doubt reflected memories of that existential threat.[10] And so, when communist aggression began to encroach on Malaya in the 1950s, the risk was keenly felt in Canberra.

This was reflected in the 1951 *Strategic Concept for Defence of the ANZAM Region*. As the Defence Committee then noted, Malaya was 'the one country which could be attacked from land and which also provided a land route from Asia to the arc of islands to the north of Australia'.[11] The defence of Malaya would ensure control over the air and sea routes to the south and deny any communist access to naval and air bases. Conversely, its loss could result in 'Australia being subject to air attacks'.[12] Unsurprisingly, when the British Government began to press for a direct Australian commitment to the Malayan Emergency in April 1950, the request was received sympathetically. The Menzies government had already determined that it would withdraw Australia's contingent from the British Commonwealth Occupation Force in Japan, and since the Korean war had not yet broken out (though it was imminent), the request was well-timed to receive a favourable reply.[13]

Responding to the British Government's request for RAAF transport and bomber aircraft for service in the Far East, Australia's Defence Committee, after consideration on 27 April 1950, advised that an air transport squadron, as well as a small detachment of four Avro Lincoln bomber aircraft, 'could be sent almost at once'.[14] In this, and later, commitments, the Australian Government applied a prudent, if not to say parsimonious, calculus. Australian forces 'were used as bargaining chips, conceded reluctantly, in response to considerable pressure, on the understanding that their military significance was less important than their diplomatic value'.[15] Thus, Australia's long-running involvement in the Malayan Emergency commenced in 1950. Although the Emergency was arguably only a small-scale conflict in strictly military terms, the stakes were high considering Malaya's significance to Australia's security concerns in the region.[16] Moreover, as a counter-insurgency campaign, the conflict saw Australian forces contending with unfamiliar and challenging circumstances within the British led-coalition.

An unconventional war

From the outset of the Emergency in 1948, the British Government in Malaya, under High Commissioner Sir Henry Gurney, considered the conflict to be a counter-insurgency action directed primarily at defeating communist ideology. To that end, British strategy emphasised the need for 'armed support for a political war, not political support for an army war'. The civilian government, not the armed forces, were to control the conflict.[17] Coalition armed forces were to help the government restore law and order, as distinct from the more conventional role of defeating the enemy militarily.

That said, by the beginning of 1950, the frequency and intensity of attacks by communist terrorists (CTs) against the government prompted High Commissioner Gurney to seek the appointment of a military Director of Operations. On

21 March that year, Lieutenant General Sir Harold Briggs was appointed to the role. Wasting no time, Briggs set policy as soon as he arrived in theatre: the government would seek to eliminate the whole communist organisation in Malaya. Importantly though, Briggs thoroughly understood the need to build popular support for his strategy. He was adamant there could be no final victory 'until the population at large had been convinced that their best future lay with the defeat of the communists'.[18]

Bolstering the security environment to counter communist influence would be pivotal; it would reassure the population of the government's intentions and capabilities and help collect intelligence useful for countering communist influence. However, building that security would be contingent on the government's ability to protect the population while effectively delivering services throughout Malaya. Briggs advocated for a major increase in the size of the police force, particularly the intelligence arm of the police, 'which had to develop far more extensive and effective sources of information'. The counter-insurgency methods he initiated were later enhanced by his successor General Sir Gerald Templar, and focused on intensive patrols and ambushes, intelligence collection, and limiting communist supply routes to starve the CTs of popular support.[19]

Thus, the context for applying air power in Malaya was a civil-military counter-insurgency campaign with a central emphasis on defeating communist ideology and geared towards asserting the government's ascendancy. Coming so soon after the Second World War, during which military strategy had taken precedence, this new context was unfamiliar to air power planners. An element of uncertainty on how best to apply air power was expected when there was no well-defined strategy for its application. In 1950, UK Secretary for War John Strachey encapsulated the ancillary role he intended air power to play, considering the imperative of winning over the civilian population.[20] 'It was never, and never could have been, a central plank of the offensive counter-insurgency military strategy'.[21]

Air operations in Malaya: the counter-insurgency context

Even as the UK was developing a tough-minded, well-executed counter-insurgency strategy, appreciation of this new type of warfare within Australian Government circles, particularly concerning the application of air power, was still in its infancy. The methods used and subsequent debates about the efficacy of air strikes in Malaya were indicative of Australia's unfamiliarity with counter-insurgency methods at the time. The RAAF forces deployed to the Malayan Emergency supported two distinct air power roles: air strike and air transport. The latter, supported by C-47 Dakota transport aircraft of 38 Squadron RAAF, was widely accepted as an important enabler of the ground campaign. According to Air Vice Marshal Sir Francis Mellersh, Air Officer Commanding Air Headquarters Malaya from 1949 to 1951:

> The most important role of the Royal Air Force in this campaign is that of air supply. Without air supply the depth of penetration of security patrols into the jungle would be limited by the amount of food and other necessities the

men can carry on their backs. This would limit their operations largely to the jungle fringes, and the insurgents would know they were free from attack in the deeper jungle.[22]

Yet despite the importance of the air transport operations, they were essentially tactical in nature, acting as an enabler of ground force operations. As such, they influenced air power strategy less than technology. This chapter, therefore, focuses primarily on the conduct of air strike operations in Malaya.

Air strike operations in Malaya: technologies and methods

Many limitations and constraints affected the application of air power in the Malayan Emergency. Firstly, the political context of the civilian-led counter-insurgency granted air power only a peripheral role. As Andrew Mumford and Caroline Kennedy-Pipe have noted, the British Government's imperative of winning the hearts and minds of the Malayan civilian population 'ensured that an aggressive and highly visible aerial presence was not viable'.[23] At the same time, there were other major impediments to offensive air strike operations, including the relative scarcity of air force assets, the technological limitations of the available air strike platforms and their supporting systems, and difficulties in collecting and distributing target intelligence. Air strike operations in Malaya were conducted with only a small air force, employing technologically simple aircraft and systems. The toughness of the operating environment compounded this constraint: the thick jungle canopy made navigation and target location remarkably difficult, and monsoonal weather exacerbated the many operational challenges and put significant strain on Allied aircrews.

One of the most significant limitations on the effectiveness of air strike operations in Malaya was the small number of available aircraft. The order of battle of the Commonwealth air forces deployed throughout the Emergency was minuscule compared to the subsequent conflict in Vietnam. For the first two years of the Emergency, the force comprised only 29 aircraft—the equivalent of three and a half squadrons—and included a mix of fighters (Spitfires and Beaufighters) and flying boats (Sunderlands).[24] Later, the force grew somewhat, with more capable aircraft available in theatre, including Hornet, Brigand and Venom fighters, and Lincoln and Canberra bombers. Even so, there was never really a sufficiently strong air force to mount a sustained campaign of any sort.[25]

Within the ebb and flow of the small Allied air contingent, perhaps the most enduring force component was provided by the Avro Lincoln bombers of No. 1 Squadron RAAF.[26] While the UK sent a small number of these aircraft to Malaya in 1950, the bombers and their aircrews were frequently rotated back to England. Since Royal Air Force (RAF) crews generally did not stay in theatre long enough to adapt fully to Malaya's peculiar operational conditions, there was little or no continuity of experience to be shared or developed. Conversely, Australia's No. 1 Squadron remained deployed to Malaya for eight years, which enabled a commensurate accumulation of knowledge and skill. The RAAF's Lincolns not only provided a noteworthy increase to the available striking force, but they also generated continuity in the air campaign, which was particularly valuable considering the challenges caused by the harsh operating environment.

Compared to later jet aircraft, the Lincoln's durability and mechanical simplicity made it reliable and well-suited to the theatre. An enlarged and improved version of its more famous Lancaster predecessor, the Lincoln carried sufficient fuel to enable it to range from its operating base in Tengah, Singapore, over the entire Malay peninsula and loiter for extended periods over target areas, with an endurance of over eleven hours. It carried a heavy bomb load, typically 14 1000-pound bombs. Sometimes, a single 4000-pound 'blockbuster' bomb was carried to clear enough jungle to allow a helicopter to land; however, these bombs were old wartime stock, with unreliable pressure fuses that tended to explode above the canopy of foliage, blowing the tops off the trees but leaving the trunks intact. The Lincoln was also equipped with two 20-millimetre cannons in the upper turret and four .50 calibre machine guns in the tail. With a full ammunition load, it became standard practice to strafe areas that had just been bombed, using all six guns. As this could be conducted from low level, the tactic was considered a quite effective deterrent against communist insurgents.[27] Moreover, the Lincoln handled well at the slow speeds required for operations at low altitudes.

The RAAF developed a range of techniques to contend with the many operational challenges they encountered, particularly concerning the difficulties involved in navigation over the featureless jungle. The Australian Mk.30 variant of the Lincoln was only equipped with basic navigation equipment, including H2S Mk.3 radar and a rudimentary wartime pulse radar beacon called Lucero. Marconi TR 1154/1155 radios were initially carried, but these were enhanced in 1955 with Loran navigation aids and radio compasses.[28] Regardless, these systems were of limited utility in Malaya, so although it was common practice to navigate visually to a briefed target, even this required superior navigation skills. Some experienced crews could take navigational fixes by observing differences in the colour of the jungle they had frequently overflown. Later in the Emergency, greater sophistication in targeting was achieved by cooperating with forward air controllers (FACs)—experienced observation pilots in Auster observation aircraft who marked targets and guided crews on bombing runs.[29] This innovation was so effective that it later saw widespread application in Vietnam.[30]

But even as aircrews became accustomed to these methods, they recognised the need for greater refinement. Bombers circling at low altitudes to identify targets tended to remove the element of surprise. Hence, an improved method involved the FAC marking the start of an attack run with a 'gate' of smoke markers or flares, over which the Lincolns would fly on an exact heading to then drop their bombs after a pre-determined time. Ground forces applied similar methods by marking a target run-in gate with various signals. This method was further refined after 1956 when a mobile ground radar unit was deployed to provide radar vectors to guide attacking aircraft to their targets accurately; it also made bombing possible in all weather conditions and at night. Indeed, the RAAF's expertise in these methods eventually meant that No. 1 Squadron was the only unit in Malaya tasked with bombing at night.[31]

Importantly though, communist insurgents quickly appreciated the need to limit their exposure to the threat of air attack. Early in the Emergency, they began to take precautions to render themselves less easily identifiable from the air, including camouflage and the dispersal of their forces and camps. Such precautions contributed to a broader problem air forces in Malaya faced regarding intelligence collection and dissemination. Specific targets could only be struck if their precise location was known, so accurate and timely intelligence was essential to enable precision strike operations. But the collection and timely distribution of such intelligence in Malaya proved remarkably difficult.[32]

While photographic reconnaissance missions were flown, they faced the same limitations as strike aircraft: the jungle and the weather. Hence, the main source of information became human intelligence, gathered from insurgents who had either been captured or surrendered.[33] However, captives could seldom read maps, nor could they understand aerial photographs, so they could not be relied on to identify the location of their fellow insurgents accurately.[34] These difficulties in targeting were further compounded by the frequent inability of Allied ground forces to specify their own position with any accuracy.[35] Moreover, timely dissemination of intelligence, regardless of its validity, was a constant challenge.[36] Consequently, as early as 1950, the Director of Operations in Malaya, responding to reports from security forces about the challenges involved in targeting the communist insurgents, determined that area-bombing would be more effective than precision attacks.[37] For the remainder of the conflict, offensive air operations emphasised area bombing rather than close air support or pinpoint strikes.

The decision to switch to area bombing is a telling example of the limits of military technology affecting air power strategy. The challenges of navigation, visual targeting (with no means of seeing through the jungle canopy), and intelligence collection and dissemination were all consequences of the limitations of available technology. Taken together, they rendered precision strike impracticable. The only real alternative was to fall back on a method familiar to the RAF from its recent experience in the Second World War—area bombing. Applying that method to a small-scale counter-insurgency campaign against a developing economy was far removed from the circumstances that originally underpinned the use of area bombing. As the name implies, area bombing is the opposite of precision strikes. It was originally applied on a massive scale against Germany and Japan in the Second World War. The RAF's night-time campaign against Germany, which involved thousands of bomber raids, sought to damage the enemy's industry, economy, and civilian morale to bring about the grand strategic goal of Nazi Germany's surrender. The huge number of missions flown and aircrews lost were seen by Bomber Command's chief, Air Marshal Arthur Harris, as a necessary investment against the value and magnitude of the targets—which is to say, entire German cities.[38] The same was true of the US strategic fire-bombing campaign against Japan.[39]

But how did any of this apply in Malaya? As it lacked any significant industrial or economic infrastructure, the communist insurgency could scarcely be regarded as a strategic target system. In any case, the small size of the Allied bombing force available meant that area strikes could only achieve a limited concentration of destructive force. RAAF operations typically involved four to six aircraft, each with its bomb load of 14 1000-pound bombs. Even accounting for the laudable proficiency displayed by the aircrews in conducting strike operations, the navigational, targeting and intelligence difficulties they encountered meant that their bombloads stood little hope of achieving any direct military effect. Little wonder that questions arose about the efficacy of the air strike campaign even while it was being conducted.

Questions over the efficacy of air strike operations

The area bombing campaign was quick to attract the attention of critics questioning the justification of its continuation. An increase in the level of air attacks from mid-1950 'led to pressure on the authorities in Malaya to justify the commitment of the material and financial resources involved'.[40] Concerned officials wanted to see 'bang for the buck', and not unreasonably, they looked for concrete, measurable outcomes that benefitted the counter-insurgency campaign— namely, enemy targets destroyed or insurgent fighters killed. But while the numbers of bombs dropped or hours flown could be measured relatively easily, it was far harder to gauge the effects of air strikes on the enemy. The inaccessibility of target areas to Allied ground forces meant that post-strike analyses took place long after a given strike had been conducted—if they were made at all. Combined with the insurgents' tendency to quickly withdraw from areas subject to air attack, post-strike 'bomb damage assessments' were seldom fruitful. From the outset, offensive air operations drew criticism from many quarters, including politicians or military planners and even those responsible for flying the missions.[41]

Several post-war analyses emphasise the low number of insurgents killed by air strikes. When measured against the grim yardstick of the body count, the results do seem underwhelming. Between 1950 and 1958, No. 1 Squadron's Lincolns dropped 85 per cent of the 35,000 tons of bombs used during the campaign, but from almost 4000 sorties, the squadron killed only 23 guerrillas.[42] Considering the number of sorties flown and the tonnage of bombs dropped, this works out at a 'particularly laboured ordnance-to-kill ratio'.[43] In a contemporary assessment, noteworthy counter-insurgency expert Brigadier General Clutterbuck argued that

> hundreds of tons of bombs were dropped on the jungle every month, but they probably killed fewer than half a dozen guerrillas a year, more by accident than design. Such senseless swiping induced a feeling of contempt for the power of modern weapons, and the enemy made full use of this contempt in their propaganda.[44]

Such views were commonplace. A RAND Corporation assessment conducted shortly after the Emergency judged that in terms of firepower delivered by air onto terrorist targets, 'the meagre results were out of all proportion to the extensive effort engaged'.[45] Even the commander-in-chief of the Far East Air Force in 1950, Air Marshal Francis Fogarty, conceded that the critics' case was strong. When asked whether air strikes were being used for lack of alternative, he wrote:

> The answer is frankly yes. If a better and more economical alternative could be found it would certainly be adopted. But at present, the most effective method and frequently the only one we can adopt to kill and harass the enemy and prevent him from regrouping in the jungle is by striking at him from the air.[46]

In the absence of kills, government authorities in Malaya sought other metrics that might provide assurances of success. In a 1952 memorandum, the Air Officer Commanding Air Headquarters Malaya, Air Vice Marshal Sir George Mills, admitted that 'it has always been recognised that the chances of direct kills are very slight'. His view was that the effect of air strikes could only be evaluated within the context of ground force operations that air power had supported. However, measuring the extent to which air strikes had had such an effect was not straightforward. On an admittedly arbitrary basis, Mills determined that air power could be said to have contributed to ground force operations if an air attack or supply drop had taken place 'within ten miles of a kill, capture or surrender that had occurred no more than 28 days after the use of air power.' While this logic might appear specious in retrospect, it nevertheless prevailed and remained the basis for measuring the effectiveness of the air offensive for the duration of the Emergency.[47]

Notwithstanding widespread thinking that linked kill ratios to effectiveness, there were other, more nuanced views. Most significantly, the Director of Operations in Malaya up to 1952, Lieutenant General Briggs, strongly supported the air strike campaign, arguing that 'offensive air support plays a very vital role in the main object of the security forces, namely, the destruction of bandit morale and the increasing of the morale of the civil population'.[48] Contradicting those who equated metrics with effectiveness, Briggs insisted that the infliction of casualties on the enemy was incidental. The objectives of the air offensive were much wider than crudely recording the number of kills.[49]

A high mission-to-kill ratio was never really a prospect, given the operating environment. But more importantly, the use of such metrics to evaluate the effectiveness of air strikes in contributing to the counter-insurgency effort was in many ways misleading. Such figures offered little insight into whether the wider aims of the security forces were being met, and in any case, the body-count numbers were often of dubious accuracy. There were many occasions when the inaccessibility of strike zones to Allied ground forces rendered such metrics implausible. In any case, as Briggs himself intimated, offensive air operations were conducted with wider objectives in mind, including the dispersal of insurgent groups, driving them into ambushes, or moving them into country suitable for ground operations, and harassment operations to lower insurgent morale.[50]

This latter school of thought reflects a more rigorous appreciation of the relationship between the application of military power and wider political, ideological, and coercive dimensions of counter-insurgency warfare. As espoused by Clutterbuck and the like, the crude equation of kill rates with military effectiveness suggests only a linear logic that assumes that the means of bringing about a change in the enemy's ideology or political thinking is by inflicting greater (or more accurate) violence. But as many latter-day analyses have suggested, irregular warfare is more concerned with the political struggle for the allegiance and support of the population. Indeed, as Richard Newton notes:

> The ability to change behaviours through the parallel mechanisms of influence and coercion [are pivotal]. Persuading the opponent that political objectives will not be attained, rather than threatening punishment unless combat actions cease, provides the critical leverage for coercion in irregular warfare.[51]

This logic is entirely germane to the Malayan Emergency. Briggs and Templar well understood that coercing the local population into supporting the government would be counterproductive and that military power, when applied, needed to support the wider political goals of the administration, in particular, to drive home the message both to the insurgents and the local populace that the government model offered the best prospect of long-term security and stability.[52] While a deeper analysis of the relationship between military force and coercion is beyond the scope of this chapter, such issues nevertheless remain important in considering the relationship between military technology and air power strategy. Considering subsequent experience in Vietnam, where body-count metrics and the widely held belief in military technology as the vehicle to strategic outcomes were to have such adverse effects, the experience of Malaya appears not to have been fully appreciated.[53]

Conclusion: lessons learnt?

Rather than addressing the more difficult political and ideological dimensions of counter-insurgency strategy, Allied air forces in Vietnam concentrated on technologies to overcome the difficulties first encountered in Malaya. Before long, precision-guided munitions, accurate navigation systems, and targeting sensors that could see through the darkness, bad weather and jungle terrain combined to ensure that air power could reliably destroy almost any target. But as impressive as all this technology became, it still failed to get to the more fundamental questions at the heart of irregular warfare. Rather than examining how, or even whether, the application of military force might convince insurgents to admit the superiority of Allied ideology, the technology-led approach simply presumed that more accurately delivered weapons would bring about the desired outcomes.

Soon after the Vietnam War, famed Viet Cong leader Ho Chi Minh revealed the dubious merit of this deterministic thinking. When told by a senior US Army officer that Vietnamese forces had never beaten US troops on the battlefield, Ho replied: 'that may be so, but it is also irrelevant'.[54] If Malaya may be regarded as one of only a few successful 20th-century counter-insurgency campaigns, then its lessons concerning the limited validity of technology-enabled strike operations—and even more so the questionable value of body-count metrics—should have been more thoroughly taken on board. Vietnam and later wars tend to suggest that those lessons went unlearnt.

Notes

1 See: Peter Edwards and Geoffrey Pemberton, *Crises and Commitments: The Politics and Diplomacy of Australia's Involvement in Southeast Asian Conflicts, 1948–1985* (Sydney: Allen and Unwin, 1992).
2 Ibid.
3 Peter Edwards, *Learning from History: Some Strategic Lessons from the 'Forward Defence'* (Canberra: Australian Strategic Policy Institute, 2015), 6.
4 Alan Stephens, *RAAF Policy Plans and Doctrine 1946–1971* (Canberra: RAAF Air Power Studies Centre, 1995), 14.
5 Robert S. McNamara, *In Retrospect: The Tragedy and Lessons of Vietnam* (New York: Times Books, 1995).
6 Edwards, *Learning from History*, 7.
7 Stephens, *RAAF Policy Plans and Doctrine 1946–1971*, 14.
8 Stephan Frühling, *A History of Australian Strategic Policy since 1945* (Canberra: Defence Publishing Service, 2009), 207.
9 Ibid., 206.
10 Edwards and Pemberton, *Crises and Commitments*, 61.
11 Alan Stephens, *Power Plus Attitude: Ideas, Strategy and Doctrine in the Royal Australian Air Force 1921–1991* (Canberra: RAAF Air Power Studies Centre, 1992), 120.
12 Ibid., 120.
13 Robert O'Neill, *Australia in the Korean War 1950–1953: Volume I – Strategy and Diplomacy* (Canberra: Australian War Memorial and the Australian Government Publishing Service, 1981), 37.
14 Peter Dennis and Jeffrey Grey, *Emergency and Confrontation: Australian Military Operations in Malaya and Borneo 1950–1966* (Sydney: Allen and Unwin, 1996), 37.
15 Edwards, *Learning from History*, 10.
16 Dennis and Grey, *Emergency and Confrontation*, 3.
17 Alan Stephens, *Going Solo: The Royal Australian Air Force 1946–1971* (Canberra: Australian Government Publishing Service, 1995), 245.
18 Dennis and Grey, *Emergency and Confrontation*, 13–14.
19 Ibid., 15, 20–21.
20 Ibid., 20–21.
21 Andrew Mumford, "Unnecessary or Unsung? The Utilisation of Airpower in Britain's Colonial Counterinsurgencies," *Small Wars & Insurgencies* 20, no. 3–4 (December 2009): 641.
22 Francis Mellersh, "The Campaign Against the Terrorists in Malaya. Lecture at the Royal United Services Institute, 7 March 1951," *RUSI Journal* 96, no. 583 (1951): 401–15, 408.
23 Andrew Mumford and Caroline Kennedy-Pipe, "Unnecessary or Unsung? The Strategic Role of Air Power in Britain's Colonial Counter-Insurgencies," in *Air Power,*

Insurgency and the 'War on Terror', ed. Joel Hayward (Cranwell: Royal Air Force Centre for Air Power Studies, 2009), 71.

24 Dennis and Grey, *Emergency and Confrontation*, 33.

25 Malcolm Postgate, *Operation Firedog: Air Support in the Malayan Emergency 1948–1960* (London: HMSO, 1982), 165–69.

26 Postgate, *Operation Firedog*, 35.

27 Mike Garbett and Brian Goulding, *Lincoln at War, 1944–1966* (London: Ian Allan, 1979), 83–84.

28 Garbett and Goulding, *Lincoln at War*, 83–84.

29 Stephens, *Going Solo*, 250.

30 See: Gary Robert Lester, *Mosquitos to Wolves: The Evolution of the Airborne Forward Air Controller* (Maxwell Air Force Base, Alabama: Air University Press, 1997).

31 Stephens, *Going Solo*, 251.

32 Director of Operations – Malaya, *Review of the Emergency in Malaya from June 1948 to August 1957 [Bowen Report]*, A452, 1968/4248, National Archives of Australia.

33 R.W. Komer, *The Malayan Emergency in Retrospect: Organisation of a Successful Counterinsurgency Effort* (Santa Monica, CA: Rand, 1972).

34 Arthur Barondes, *The Accomplishments of Airpower in the Malayan Emergency* (Maxwell Air Force Base, AL: Concepts Division, Aerospace Studies Institute, Air University Press, 1963), 50.

35 Ibid., 50.

36 David Ucko, *Network Centric Operations Case Study: The British Approach to Low Intensity Operations* (Washington, DC: US Department of Defence Office of Force Transformation Technical Report, 2007).

37 Dennis and Grey, *Emergency and Confrontation*, 34.

38 See: Noble Frankland and Charles Webster, *The Strategic Air Offensive Against Germany 1939–45*, vols 1–4 (London: HMSO, 1961).

39 For examples, see United States Government, *Strategic Bombing Survey Summary Report (Pacific War)* (Washington, DC: US Government Printing Office, 1946).

40 Dennis and Grey, *Emergency and Confrontation*, 35.

41 Stephens, *Going Solo*, 249.

42 Ibid.

43 Sebastian Ritchie, *The RAF, Small Wars and Insurgencies: Later Colonial Operations, 1945–1975* (Cranwell, Lincolnshire: RAF Air Historical Branch, 2011), 21.

44 Dennis Drew, "Airpower in Peripheral Conflict: From the Past, the Future," in *The War in the Air 1914–1994*, ed. Alan Stephens (Maxwell Air Force Base, AL: Air University Press, 1991), 265.

45 A. Petersen, G. Reinhardt, and E. Conger, eds., *Symposium on the Role of Airpower in Counterinsurgency and Unconventional Warfare: The Malayan Emergency* (Santa Monica, California: Rand Corporation, 1963).

46 Dennis and Grey, *Emergency and Confrontation*, 37.

47 Ibid., 39.

48 Ibid., 38.

49 Stephens, *Going Solo*, 250.

50 Ritchie, *The RAF, Small Wars and Insurgencies*, 24–25.

51 Richard Newton, "Air Power, Coercion, And . . . Irregular Warfare?" *RAF Air Power Review* 13, no. 2 (Summer 2010): 6.

52 McNamara, *In Retrospect*, 6.

53 Robert Pape, *Bombing to Win. Air Power and Coercion in War* (Ithaca: Cornell University Press, 1996).

54 Harry G. Summers, "Interview with General Frederick C. Weyand about the American Troops Who Fought in the Vietnam War," *Vietnam Magazine (Leesburg, Virginia)* (Summer 1988), reproduced at: www.historynet.com/interview-with-general-frederick-c-weyand-about-the-american-troops-who-fought-in-the-vietnam-war/.

References

Archival sources

National Archives of Australia
A452: Department of Territories, Correspondence files, 1901–1990.

Government publications
United States Government. *Strategic Bombing Survey Summary Report (Pacific War)*. Washington, DC: US Government Printing Office, 1946.

Published sources
Barondes, Arthur. *The Accomplishments of Airpower in the Malayan Emergency*. Maxwell Air Force Base, AL: Concepts Division, Aerospace Studies Institute, Air University Press, 1963.

Dennis, Peter, and Jeffrey Grey. *The Official History of Australia's Involvement in Southeast Asian Conflicts 1948–1975. Emergency and Confrontation: Australian Military Operations in Malaya and Borneo 1950–1966*. Sydney: Allen and Unwin, 1996.

Drew, Dennis. "Airpower in Peripheral Conflict: From the Past, the Future." In *The War in the Air 1914–1994*, edited by Alan Stephens, 257–300. Maxwell Air Force Base, AL: Air University Press, 1991.

Edwards, Peter. *Learning from History: Some Strategic Lessons from the 'Forward Defence'*. Canberra: Australian Strategic Policy Institute, 2015.

Edwards, Peter, and Geoffrey Pemberton. *The Official History of Australia's Involvement in Southeast Asian Conflicts 1948–1975. Crises and Commitments: The Politics and Diplomacy of Australia's involvement in Southeast Asian Conflicts 1948–1965*. Sydney: Allen and Unwin, 1992.

Frankland, Noble, and Charles Webster. *History of the Second World War, United Kingdom Military Series, The Strategic Air Offensive against Germany*. 4 vols. London: HMSO, 1961.

Frühling, Stephan. *A History of Australian Strategic Policy Since 1945*. Canberra: Defence Publishing Service, 2009.

Garbett, Mike, and Brian Goulding. *Lincoln at War 1944–1966*. London: Ian Allan, 1979.

Komer, R. W. *The Malayan Emergency in Retrospect: Organisation of a Successful Counterinsurgency Effort*. Santa Monica, CA: Rand, 1972.

Lester, Gary Robert. *Mosquitos to Wolves: The Evolution of the Airborne Forward Air Controller*. Maxwell Air Force Base, AL: Air University Press, 1997.

McNamara, Robert S. *In Retrospect: The Tragedy and Lessons of Vietnam*. New York: Times Books, 1995.

Mellersh, Francis. "The Campaign Against the Terrorists in Malaya: Lecture at the Royal United Services Institute." March 1951 *RUSI Journal* 96, no. 583 (7 March 1951): 401–15.

Mumford, Andrew. "Unnecessary or Unsung? The Utilisation of Airpower in Britain's Colonial Counterinsurgencies." *Small Wars & Insurgencies* 20, no. 3–4 (2009): 636–55.

Mumford, Andrew, and Caroline Kennedy-Pipe. "Unnecessary or Unsung? The Strategic Role of Air Power in Britain's Colonial Counter-Insurgencies." In *Air Power, Insurgency* and the 'War on Terror', edited by Joel Hayward, 67–79. Cranwell: Royal Air Force Centre for Air Power Studies, 2009.

Newton, Richard. "Air Power, Coercion, and . . . Irregular Warfare?" *RAF Air Power Review* 13, no. 2 (Summer 2010): 1–20.

O'Neill, Robert. *Australia in the Korean War 1950–53: Volume I – Strategy and Diplomacy*. Canberra: Australian War Memorial and the Australian Government Publishing Service, 1981.

Pape, Robert. *Bombing to Win: Air Power and Coercion in War*. Ithaca: Cornell University Press, 1996.

Petersen, A., G. Reinhardt and E. Conger, eds. *Symposium on the Role of Airpower in Counterinsurgency and Unconventional Warfare: The Malayan Emergency*. Santa Monica, CA: Rand Corporation, 1963.

Postgate, Malcolm. *Operation Firedog: Air Support in the Malayan Emergency 1948–1960*. London: HMSO, 1982.

Ritchie, Sebastian. *The RAF, Small Wars and Insurgencies: Later Colonial Operations, 1945–1975*. Cranwell, Lincolnshire: RAF Air Historical Branch, 2011.

Stephens, Alan. *Power Plus Attitude: Ideas, Strategy and Doctrine in the Royal Air Australian Air Force 1921–1991*. Canberra: RAAF Air Power Studies Centre, 1992.

———. *Going Solo: The Royal Australian Air Force 1946–1971*. Canberra: Australian Government Publishing Service, 1995a.

———. *RAAF Policy, Plans and Doctrine*. Canberra: RAAF Air Power Studies Centre, 1995b.

Ucko, David. *Network Centric Operations Case Study: The British Approach to Low Intensity Operations*. Washington, DC: US Department of Defence Office of Force Transformation Technical Report, 2007.

Part Two
Identity and Culture

5 The importance of asking why the Royal Australian Air Force Exists[1]

Jason Begley and Travis Hallen

Why does air power require an independent service? This question has proven enduring for air power professionals, emerging periodically in many Western militaries, particularly during debates on future force structure in which the competing budgetary pressures that democratic governments face inevitably play a key role. Historically, the Royal Australian Air Force (RAAF) endured regular criticism during the Great Depression, when Australian Army and Royal Australian Navy officers sought to bring it under their command and challenged its existence as a separate service.[2] The primary justifications supporting an independent air force originate in the earliest days of military aviation. Leading air power theorists and advocates of the interwar period such as Giulio Douhet and William 'Billy' Mitchell argued passionately, and at times with foolhardy conviction, that the nature and potential of operations in and from the air necessitated the creation of a third service that was co-equal with the navy and the army.[3] Though the case for an independent air force varied based on each country's cultural, political, organisational, and historical circumstances, a common thread was woven through each: the belief that only a service commanded by an aviator could ensure the effective employment of air power. This belief that a thorough understanding and experience of air power is essential to its effective employment remains an article of faith for modern air power professionals.

An aviator's ability to optimise the employment of air power is not, however, an innate quality of those who join an air force, nor is it bestowed upon them when they don an air force uniform; it is the result of a career-long investment in education, training, and experience. Unfortunately, the RAAF has long struggled with developing and implementing an effective Professional Military Education and Training (PMET) system that promotes the professional mastery of air power within its workforce. Although the RAAF is investing in improving PMET, much work remains. One area that requires particular attention is the ability of RAAF aviators to understand and articulate a core principle of the service to which they belong—the enduring requirement for an independent air force in the Australian Defence Force.

In Australia, the rationale for the RAAF's existence is rarely discussed outside the Australian Command and Staff College (ACSC). In contrast, it is still

DOI: 10.4324/9781003230656-8

regularly debated in other countries, such as the United States.[4] Although this question is rarely asked, it must play a prominent role in the education and mentoring of Australian aviators. Appreciating the function of an air force is more than an exercise in justifying the continued existence of a separate service; it is a subject that invites deep consideration of the importance of air power and the role of the RAAF in Australia's evolving geopolitical environment.

This chapter evolved from the premise that engaging in a discussion on the raison d'être of the RAAF would create the opportunity for commanders to engage the aviators in their charge and develop within them a better understanding of the air power profession. By understanding their profession first, RAAF personnel would be better able to understand and contribute to discussions relating to defence and strategic matters and recognise when and how they might innovate. Based on the idea that critical thinking and innovation are important aspects of an aviator's professional development, a decision was made not to approach the topic by writing a traditional academic paper but rather to use a Socratic-dialogue type discussion between a squadron commanding officer (CO) and his executive officer (XO).

This format was chosen for three reasons. Firstly, it highlighted the importance of mentoring in developing the minds of junior air power professionals. The Socratic Method aims to develop independent insight by focusing not on what to think but on *how* to think. Air power mastery and PMET in the RAAF have traditionally focused on the former approach, a mode of thinking that is less conducive to the innovative thought necessary to ensure the RAAF's continued strategic relevance. By adopting a Socratic dialogue, we intend to highlight both the benefits of mentoring in developing the mind of the aviator and the responsibility of commanders to do more than simply regurgitate doctrine and policy to their people.

Secondly, the Socratic dialogue provides a way to engage an audience more effectively in exploring this topic. A to-and-fro discussion about the RAAF's reason for existence allowed a more rapid connection and transition between concepts than would be possible using a more traditional structure. Socratic dialogue allows coverage of this broad topic in sufficient depth to draw out key points and stimulate readers' thinking. Finally, the Socratic discourse is non-traditional. The willingness of the early leaders of the RAAF to challenge the status quo by pushing for an independent air service remains a core foundation myth and a source of pride for the organisation as an iconoclastic service. In this context, opting for a non-traditional format in the interests of best effect is arguably appropriate.

The question of the RAAF's continued independence provides an excellent vehicle to explore the broader questions of air power's role in the Australian military and the importance of educating our aviators to enhance the future RAAF. The following discussion is less about answering the so-called air power question than emphasising the critical importance of thinking and talking about it. This point is neglected in the current educational system and is something that must change.

The Dialogue

XO: Sir, a minute of your time?

CO: Certainly, XO, come in.

XO: When I got out of the simulator yesterday, I found an email congratulating me on my selection for Command and Staff Course.

CO: Congratulations. Staff College is a key step in your development to fulfil future senior leadership roles in Air Force. You must be quite happy.

XO: Well, that's the thing. I know it's competitive, so I'm pleased to be selected, but to be honest, I don't think I'm ready for it. My career so far is flying or thinking about flying; I'm not ready to talk about Clausewitz and strategy. So, I was hoping for some advice on what to expect and how to prepare myself best.

CO: Your concerns are justified; Staff College is a step up in professional mastery for people like us who've spent fifteen to twenty years of their career thinking solely about flying operations, so asking why we do it and how that fits into broader national security considerations can seem daunting. So, how Staff College forces us to look beyond the kit and more to the key concepts is quite important.

XO: I think I get what you mean. I've seen how the junior officers view the PMET scheme as distracting them from their 'real job' as aircrew.

CO: Exactly. But I wouldn't equate Staff College to Clausewitz. You'll probably discuss the Dead Prussian, but there's much more to professional mastery at that level than 'war as an extension of politics'. There are many authors and ideas you'll explore, but from my experience, as an air force officer, you need to be ready for one key question. Without fail, during your time at Staff College, your army and navy course mates will ask this question, and you need to have thought carefully about how to answer it.

XO: What's that?

CO: Why does the Air Force exist?

XO: That's simple: to generate air power effects for Australia's interests.

CO: That's a nice starting point, but if it were that simple, the navy, and even the army, would not still be asking the question. You have given me a bumper-sticker slogan, not the reason for the Air Force—or the raison d'être if you will. At Staff College, answering questions like this is as much about 'how' you arrive at the answer as it is about the actual answer. Indulge me a little, and you'll see what I mean.

XO: Okay.

CO: Would you agree that air power is important in the modern battlespace?

XO: Definitely. Oceans cover two-thirds of the earth, land the other third, but 'the air covers the whole world, aircraft are able to go anywhere on the planet'.[5] Therefore, any operation on land or sea is vulnerable to attack from the air.

CO: Exactly. Douhet, one of the first air power theorists, made that point when he said: 'To conquer the command of the air means victory; to be beaten in the air means defeat and acceptance of whatever terms the enemy may be pleased to impose'.[6] But in itself, would you agree that this does not justify an air force?

XO: I think it does. If there wasn't an air force, who would provide air power? Isn't that the reason why the RAAF was created?

CO: That's a good question because it helps us frame our thinking, and you're right to an extent. In 1919 the Swinburne Commission recognised the essential need for an air organisation and that 'a single "Australian Air Corps" should be organised into two wings, to meet the needs of both the Navy and the Army'.[7] Would you agree that this seems to be more about efficiency than effect? And that it's somewhat redundant now, given Navy and Army both have their own air power assets?

XO: Well, yes.

CO: So, realistically, the Swinburne Committee's reason for an independent air force—to deliver air power more efficiently than the navy or the army—is no longer a valid argument, correct?

XO: To an extent, things have changed since then. The RAAF delivers air power that the Committee didn't imagine and that the army and the navy can't deliver.

CO: That's the most common argument offered for our existence. Group Captain Edgeley described it using the term 'the justification cycle'. He found that air forces have tended to justify their existence based on 'the efficiency and increased effectiveness of having air assets concentrated under one Service, and, towards the end of World War I, on an ability of air power to undertake independent action'.[8] So, let's test that assumption. What aspects of air power can only be provided by an air force?

XO: Well, that's fairly straightforward: the air power contributions.

CO: By that, you mean control of the air, intelligence, surveillance and reconnaissance (ISR), air mobility, and strike?

XO: Yes. That's the heart of doctrine and the basis of PMET.

CO: I'm pleased that you're familiar with our doctrine. But to answer our question, you need to explain how the air power contributions translate into Air Force's *raison d'être*.

XO: Well, the best place to start is control of the air—that 'is the most fundamental thing we do'.[9] We maintain awareness of the air battlespace through our land-based radars and the Wedgetail and then translate that into air superiority through our air combat assets. And then there's the ground-based air and missile defence system the 2016 Defence White Paper talked about, which I am sure will be operated by the air force; it's a natural fit.[10]

CO: You're right; we'll soon definitely have a potent control of the air capability. But, what about the Air Warfare Destroyer (AWD)? Doesn't it enable control of the air through its surveillance and control capabilities and missile systems? And with both persistence and mobility, you could argue it's a more effective capability. Meanwhile, the White Paper also described the modernisation of the army's ground-based air defence (GBAD) capabilities. So, wouldn't it be more accurate to say the air force is the major, but not the only contributor, to control of the air?

XO: Okay, the AWD offers persistence, but it doesn't have the reach or responsiveness of our fighters—20 knots isn't overly mobile. Sea lanes will also constrain the AWD, and the army's GBAD is, in essence, point defence.

CO: So, sea-based control of the air gains persistence, but at the cost of true reach—valid. But you argued that the systems in its possession limit the army's control of the air. Is there a reason why the Army couldn't operate the future land-based radar and missile systems mentioned in the White Paper?

XO: Well, they wouldn't know how to operate them because their focus is on land combat.

CO: That sounds like a status quo argument, and it's not necessarily accurate. The army is currently the only Service with GBAD experience, so on that basis, they might argue they're best positioned to control any new missile system introduced into the ADF. Wouldn't their perspective also be extended if the army owned the entire GBAD capability, including all the air surveillance systems with the extension to their effective range?

XO: I suppose.

CO: Let me go a step further. Is there a reason why 1 Remote Sensor Unit's capabilities, the Jindalee Operational Radar Network and new space systems, couldn't be operated by a national agency such as the Australian Geospatial Organisation? That would seem to align with their task appropriately.

XO: Because we have all the experience and systems to manage and develop this capability. We have the expertise, so it makes sense that the capabilities remain in Air Force.

CO: Again, your argument seems to be that the air force exists because the capabilities it acquired justify its existence. Since I mentioned national agencies and space, what are your thoughts on ISR?

XO: This is another area where Air Force has critical capabilities and expertise. But I see where you're going, so I'll try and provide more solid logic. I concede that all three services have ISR capabilities to support their operations and contribute to the broader ADF and national requirements, but the air force assets bring unique effects.

CO: Good; what makes us unique is the key to the question. So, what do you mean here—are you talking sensors, aircraft, both?

XO: It's more than that. It's how the sensors and aircraft come together to enable the exploitation of the characteristics of the air. ISR is where we truly demonstrate the flexibility, reach, and perspective that operating in the air domain affords us. We can see further, shift focus more rapidly, and expand the commander's view of the battlespace more effectively than land or sea-based systems. This is what we do continuously, which means that—if you'll forgive the pun—our perspective is different from our army and navy colleagues who develop and gain mastery based on the inherent characteristics of their domain.

CO: So, you're saying that, although the other services have capabilities that fulfil some of the air power roles, there is a difference in perspective and attitude to how these are conducted that makes the air force's contribution to air power unique?

XO: Yes, that seems to make more sense. It becomes much clearer if we look at it from the strike perspective. Kosovo is a great example where they say that air power achieved a decisive outcome on its own using precision strike.

CO: Kosovo is always an interesting case study. Arguably, Operation Allied Force achieved its strategic goal in 78 days, without the need for ground forces, validating the long-held vision of early air power theorists. Kosovo changed the perspective of some avid critics of air power; for example, just before Milošević capitulated, John Keegan stated that 'rather as a Creationist Christian . . . being shown his first dinosaur bone, he might have been wrong for the past forty years.'[11] But let me ask you, who commanded that operation?

XO: I think it was an army guy.

CO: Correct. It was General Wesley Clark. So, if an army officer led air power's most decisive outcome, why do we need an independent air force?

XO: Hang on. I remember Alan Stephens saying that Clark's running of Allied Force could be described 'as little more than a disconnected series of "random acts of violence"'.[12]

CO: Well-spotted. To paraphrase General Spaatz, the first Chief of Staff of the US Air Force, while few air force generals think they can run a land campaign, most army generals think they can conduct an air campaign.[13] While Clark's efforts may have eventually achieved the outcome, there is a strong argument that this was despite Clark's involvement, not because of it.

XO: That seems a fair point. The air domain has a different dynamic, and to optimise that, you need to develop your understanding of it from experience gained throughout your career. I would say that from a strike perspective, Army sees only as far as their immediate reach, perhaps the next few days of manoeuvre. So, they would naturally target the enemy immediately in front of them or interdict reserves of logistic resupply.

CO: That's the Jominian approach at its best—looking at decisive battle as the only method to employ force as a means to an end.[14]

XO: But an aviator would look deeper, as in Operation Desert Storm. That air campaign focused on taking out the command and control nodes, the integrated air defence system, and finally, the fielded forces. It was an aviator's idea—Warden, I think—to use air power's reach and penetration far beyond the line of our troops to degrade the enemy's ability to function effectively as a force.[15]

CO: True. Warden's concentric rings became the quintessential planning tool for air power strategy. They treated the enemy as a system and took a systems approach to operate against it, but Warden wasn't the first aviator to change how we approach dealing with an enemy. Have you heard of John Boyd?

XO: He was the Korean War fighter pilot that invented the OODA (observe–orient–decide–act) loop, right?

CO: Reducing Boyd to the OODA loop is like reducing Clausewitz to war as an extension of politics: you lose an awful lot of value in the simplification. Boyd viewed strategy as a game in which we aim to diminish the adversary's ability to communicate or interact with his environment while sustaining or improving our own, such that, 'He who can handle the quickest rate of change survives.'[16] Like the best warfare theories, this approach has universal application, and the US Marine Corps is often the biggest proponent of Boyd's theories. But it reflects how the different perspective of an aviator changes how we approach strategy. Can you think of anything similar relating to air mobility?

XO: Air mobility? I know that it's a core air power contribution. Still, it's just an enabling function air force provides at the operational and strategic level, and it's hardly unique or decisive. The army does its battlefield airlift with a little help from the air force. The navy does the shipborne stuff, and we seem to rely on commercial leases more than I would like. It's probably the least compelling argument for an independent air force.

CO: What about the Berlin Airlift? Can you see another way aside from air mobility that we could have achieved the same profound strategic effect in that case?

XO: No. I don't think we consider that example anywhere near as much as we should. But now I think about it, the air mobility operation supporting the Mount Sinjar refugees in Iraq is another good and recent case of a strategic effect using air mobility.

CO: Indeed, but let's return to our core question. At the start of this discussion, you said that the RAAF exists to provide air power effects for Australia's interests. Thoughts?

XO: It's still true, but I agree there needs to be more to it to make a compelling argument. I get that the other services also produce air power effects, but we've identified a specific context to this. There's an attitude, a perspective, almost a philosophy that underpins the way RAAF aviators approach the promotion of Australia's national interest, and that's the result of a career gaining experience in developing and operating capabilities that exploit the characteristics of the air domain.

CO: So, you're saying that aviators have an air-minded way of thinking which affords them the perspective to provide a genuinely unique air power contribution to the ADF?

XO: Pretty much.

CO: So why doesn't this apply to the army and the navy aircrew?

XO: Well, they don't devote a career to it in the way we do in the RAAF. Army and naval helicopter pilots will see their role within the context of the land or sea domain, so they use air power as a tool to deliver land and sea power. It's a subtle difference, but I think an important one. And it would be impossible for them to become experts in air power in the way of our air-minded aviators without detracting from their progression within their services.

CO: Impossible is a fairly absolute term.

XO: Not impossible then, but inefficient. The ADF needs to develop combat mastery in all the domains that we operate in, but the characteristics of each influence how they can best be exploited. So, we need the army focused on land schemes of manoeuvre, and the navy focused on maritime warfare. Their aircrew will develop an understanding of how the air domain influences operations in the other two, letting them optimise their part of air support to land and sea. But they won't be professional masters of operating in the air domain independently when the situation calls for it.

CO: And how will we know when the situation calls for it?

XO: I think that's the final piece. A true air power professional considers all the options that air power can provide, including stating when air power may not be a viable option at all. It is vital to developing people that can coherently and logically provide this advice in a non-parochial manner to support joint operations. I think this may have been what a number of Chiefs of Air Force meant when they've said that 'Air Force is fundamentally about our people.'[17]

CO: That might not be the specific context, but it aligns with his intent and is central to our question. And so, do you think you've formed a defensible position on why the RAAF exists?

XO: To be honest, I don't think so. The efficiency argument is no longer valid, as all three services now have an air power capability. Although there are subtle differences, it's the same for the air power roles, and the operational context is important. So, I would have to say that the closest answer I have is that the air force provides a framework to develop leaders that understand air power and how best to employ it, which in turn provides the ADF and the government with a similarly unique perspective to complement that provided by the other services.

CO: It would seem extravagant to maintain a service simply to train thinkers— isn't that what universities should do?

XO: Well then, I guess we haven't found a good answer—maybe this was a waste of time.

CO: On the contrary, what you've achieved is far more important than the simple satisfaction of a glib one-line explanation for why we have an air force. By exploring the question and challenging each line of reasoning, you've realised that there is more to this topic than meets the eye initially—and hopefully, this will change the way you approach topics like this in the future. As I said initially, it is not about what to think but how to think. The air force is going through a major transition in every facet of the service. To face the challenges that will arise, we need to be ready to challenge existing paradigms rather than regurgitate dogma. Remember, legacy thinking will only deliver legacy outcomes.

XO: Well, from that perspective, I understand how much I need to learn about approaching these questions and am better equipped for it.

CO: Good—innovative thinking begins with being willing to ask both 'why?' and 'why not?'

Conclusion

Although this dialogue may not have arrived at a definitive answer to why the air force should exist as an independent service, the effort is not wasted. As with Plato's original Socratic dialogues, it is not the answer at the end that matters but the intellectual journey taken to get there. Conversations like the one discussed here should be occurring at all levels and across all areas of the RAAF to complement the existing PMET system. The key to developing the intellectual foundations on which the RAAF's professional mastery of air power lies in discussions such as these. Through similar processes, aviators will be best prepared to identify and direct the innovative changes to the air force required in response to the constantly evolving technological and strategic landscape within which air power is employed. Without thinking and questioning, Australia's aviators will be left only to regurgitate a long-standing dogma that, in isolation, struggles to justify the continued relevance of air power in the modern battlespace.

Notes

1 This chapter was written before the release of the most recent *Air Power Manual* in March 2022. It, therefore, reflects the state of the field at that time and uses sources as they then stood, including the 6th Edition of the *Air Power Manual*. See: Royal Australian Air Force, *Air Power Manual*, 6th ed. (Canberra: Air Power Development Centre, 2013).

2 For example, see: Summary of proceedings, "General meeting of the Council of Defence held at Commonwealth Offices, Melbourne, at 2 P.M. on Tuesday, 12 November 1929," 12 November 1929, 7–10, A5954, 908/5, National Archives of Australia, Canberra.

3 Giulio Douhet, *Command of the Air* (Tuscaloosa: University of Alabama Press, 2009), 128–29; William Mitchell, *Winged Defense: The Development and Possibilities of Modern Air Power–Economic and Military* (Tuscaloosa: University of Alabama Press, 2009), 216–17.

4 See, for example: Robert M. Farley, *Grounded: The Case for Abolishing the United States Air Force* (Lexington, KY: University of Kentucky Press, 2014).

5 Mitchell, *Winged Defense*, 3–4.

6 Douhet, *Command of the Air*, 28.

7 Ibid.

8 Stephen Edgeley, *'Out of Joint': Independent Air Forces in Democratic Cultures* (Canberra: RAAF Air Power Development Centre, 2010), 31.

9 Speech, Air Marshal Geoff Brown, "The Role of the RAAF in Australia's National Security," 26 July 2012. For an abridged version of this speech, see: Geoff Brown, "The role of the RAAF in Australia's Security," *Australian Defence Force Journal* 190 (2013): 19–29.

10 Department of Defence, *Defence White Paper – 2016* (Canberra: Commonwealth of Australia, 25 February 2016), 96.
11 Alan Stephens, *Kosovo, or the Future of War, Paper No. 77* (Canberra: RAAF Air Power Development Centre, 1999), 5.
12 Alan Stephens, "Perception, Reality, and 21st Century Strategy," in *The Art of Air Power: Proceedings of the 2010 Air Power Conference*, ed. Keith Brent (Canberra: Air Power Development Centre, 2011), 120.
13 Spaatz's actual observation related to air force generals running armies and army generals running air force. Alan Stephens, "Fifth Generation Air Power," in *Airpower Reborn: The Strategic Concepts of John Warden and John Boyd*, ed. John Andreas Olsen (Annapolis, MD: Naval Institute Press, 2015), 152.
14 Antoine-Henri Jomini was a Swiss-born soldier who served on the staff of French General Ney and Napoleon between 1805 and 1813. He is most noted for his work on strategy *The Art of War* in which he states: 'Every maxim relating to war will be good if it indicates the employment of the greatest portion of the means of action at the decisive moment and place'. See: Antoine-Henri Jomini, *The Art of War* (Mineola, NY: Dover Publications, 2007), 295.
15 John A. Warden III, "The Enemy as a System," *Airpower Journal* IX, no. 1 (Spring 1995): 49.
16 Frans Osinga, "The Enemy as a Complex Adaptive System," in *Airpower Reborn: The Strategic Concepts of John Warden and John Boyd*, ed. John Andreas Olsen (Annapolis, MD: Naval Institute Press, 2015), 64.
17 Leo Davies, "A 10 Year Plan – an Air Force Strategy," *Australian Strategic Policy Institute*, 20 July 2016, www.aspistrategist.org.au/10-year-plan-air-force-strategy/.

References

Archival Sources

National Archives of Australia

A5954: "The Shedden Collection".

Government Publications

Department of Defence. *Defence White Paper – 2016*. Canberra: Commonwealth of Australia, 25 February 2016. www.defence.gov.au/whitepaper/docs/2016-defence-white-paper.pdf.
Royal Australian Air Force. *The Air Power Manual*. 6th ed. Canberra: Air Power Development Centre, 2013.

Published Sources

Brown, Geoff. "The Role of the RAAF in Australia's Security." *Australian Defence Force Journal* 190 (2013): 19–29.
Douhet, Giulio. *Command of the Air*. Tuscaloosa: University of Alabama Press, 2009.
Edgeley, Stephen. *'Out of Joint': Independent Air Forces in Democratic Cultures*. Canberra: RAAF Air Power Development Centre, 2010.
Farley, Robert M. *Grounded: The Case for Abolishing the United States Air Force*. Lexington, KY: University of Kentucky Press, 2014.

Jomini, Antoine-Henri. *The Art of War*. Mineola, NY: Dover Publications, 2007.

Mitchell, William. *Winged Defense: The Development and Possibilities of Modern Air Power—Economic and Military*. Tuscaloosa: University of Alabama Press, 2009.

Osinga, Frans. "The Enemy as a Complex Adaptive System." In *Airpower Reborn: The Strategic Concepts of John Warden and John Boyd*, edited by John Andreas Olsen, 48–92. Annapolis, MD: Naval Institute Press, 2015.

Stephens, Alan. *Kosovo, or the Future of War, Paper No. 77*. Canberra: RAAF Air Power Development Centre, 1999.

———. "Perception, Reality, and 21st Century Strategy." In *The Art of Air Power: Proceedings of the 2010 Air Power Conference*, edited by Keith Brent, 117–35. Canberra: Air Power Development Centre, 2011.

———. "Fifth Generation Air Power." In *Airpower Reborn: The Strategic Concepts of John Warden and John Boyd*, edited by John Andreas Olsen, 128–55. Annapolis, MD: Naval Institute Press, 2015.

Warden, John A. III. "The Enemy as a System." *Airpower Journal* IX, no. 1 (Spring 1995): 40–55.

6 Identity as a gatekeeper in Western Air Forces

Jarrod Pendlebury

In this chapter, I will explore how military personnel construct and perpetuate concepts of an ideal identity at air force officer training establishments. My interest in this area has slowly developed throughout a career during which I have had little cause to question the status quo. In most scenarios, I have represented a dominant demographic: the white, Anglo-Celtic male. There have, however, been occasions that prompted me to pause and consider the essence of who we thought we were as air force members. One such occasion occurred during a deployment to the Middle East, where my squadron was conducting operations from a large United States (US) base. A broad entertainment programme was established to break the monotony of operations. This programme included visits from entertainers and groups, such as the iconic United Service Organizations (USO) shows of the Second World War. One particularly popular event brought the Hooters Calendar Girls to our base. A Hooters executive later described the success of the event:

> The show ended with a bang when UC3 sang 'Anywhere USA' accompanied by 6 flag waving Hooters Calendar Girls sporting Hooters latest Military appreciation uniform, featuring camouflage patterned shorts and a tank top with the line 'Weapons of Mass Distraction' on the back. The show, which was designed to resemble the classic Bob Hope USO tours, was described by many troops as the most fun and best attended morale show in years.[1]

In selecting a tour ensemble that declares its business to be 'selling sex appeal', the US Government made a set of generalised assumptions about the target audience.[2] More specifically, it seems the decision to book the Hooters franchise to entertain those deployed on operations was underpinned by a broad assumption of hegemonic heterosexual masculinity.

This chapter seeks to explore what feeds the cultivation of such assumptions that appear to contradict stated policy intentions to establish and nurture an inclusive and representative military.[3] In other words, I seek to go some way in explaining how conflicting identity frameworks (for this chapter, masculine

DOI: 10.4324/9781003230656-9

heteronormativity on the one hand and representative inclusivity on the other) can coexist in today's military. In doing so, I aim to uncover some solutions to the thorny problem of under-representation in the Royal Australian Air Force (RAAF) of certain visible minorities.[4] In this chapter, I approach this general question by specifically focusing on the dimension of gender.

There is a broad consensus that cultivating inclusion within an organisation realises key normative and instrumental benefits.[5] Thus, significantly low representation of certain visible minorities, such as women in defence, may result in an underperforming organisation and reflect poorly on the value of normative human rights within a group. Analysis of the shifting demographics in Australia—drawn from a comparison of Australian Bureau of Statistics and Defence Census data—points clearly to the fact that while the broader nation diversifies, the RAAF does not reflect this shift. The shrinking of its historically dominant demographic should prompt consideration about why certain visible minorities remain represented at low levels. For instance, despite focused recruitment and retention initiatives, the representation of women in the RAAF has increased by only ten per cent since 1991.[6] At this rate, it would take until 2104 to reach parity with the levels of women's participation in the Australian workforce outside the military context.[7]

To assist in exploring these questions, I conducted fieldwork which compared air force officer training in three Western democracies that explicitly seek to build gender equality in their militaries: Australia, the United Kingdom (UK), and the US. This research consisted largely of in-depth one-on-one interviews with both cadets and staff, and focus groups with cadets. This research intended to build narratives to help inform an understanding of what constitutes an 'ideal' air force officer in each of these countries and explore how basic training contributes to the constitution of this ideal. This chapter presents some findings from fieldwork at the United States Air Force Academy (USAFA) and Royal Air Force (RAF) College Cranwell in the UK.

Theoretical frameworks

This chapter draws primarily on the sociological theories of Erving Goffman and gender theorist Judith Butler to suggest that air force identity is created through two distinct, sequential processes. For Goffman, social life constitutes a 'dramaturgical' environment in which impressions are given and received. Goffman coined the word 'dramaturgical' to articulate the theatrical nature of social interaction, using the term 'presentation' to describe the process whereby a set of characteristics are presented (either cynically or genuinely) to make an impression. Importantly, Goffman suggests this process of presentation has the capability of 'moulding and producing an identity'.[8] At the USAFA, for instance, cadet induction relies heavily on the dramaturgical, with senior cadets assuming a stereotypical military instructor role akin to popular characterisations of a boot camp drill instructor. Throughout the process, the cadets are presented with an idealised identity that feeds into their development of *esprit de corps*. As one freshman cadet observed:

> Looking on the outside, somebody who's not in it . . . the things we had to do are kind of like so ridiculous that they're funny. Like the way we had to eat . . . having to take small bites . . . seven chews . . . you had to put your silverware down and then your hands back in your lap . . . and chew and swallow . . . and then another bite . . . the cadre are basically creating an environment where you had to come together and work against them . . . you know . . . or work against the system . . . And so, on one hand, that really did do a lot for bringing us together.[9]

A journalist who spent time at the USAFA observing an entire academic year articulates the abrupt, jarring nature of this shift into the dramaturgical:

> The dulcet tones of the cadre [upper-class training cadets] vanish the very instant the bus turns the corner, away from the Cadets' parents. 'Eyes forward!' a voice booms suddenly, a prelude to the world that awaits them over the next few weeks. The cloying dies and its exact opposite is born, hurling the newcomers into the freshman's life of orders and criticism, mind games, and exuberantly possessive pronouns. The change comes so unexpectedly, and so completely that it seems choreographed (It is).[10]

Judith Butler's theory of performativity helps explain the subsequent process of identity formation that occurs within the framework of this dramaturgical role-play. Butler views gender as constructed through the 'stylised repetition of acts', thereby refuting the idea that there is such a thing as an essential gender tied to one's biological sex.[11] For Butler, the continuity of such acts is crucial to establishing gender as a meaningful label. By broadening this idea of performativity outside the gender space and applying it in a military context, this chapter argues that basic military training is infused with 'stylised repetition[s] of acts' that help reinforce the presented (in a Goffmanian sense) identity. In essence, the air force identity shared by members of these air forces is a social construction infused with concepts of the classical, masculine warrior. Moreover, this identity is assumed through a process of performative acquisition.

In *Asylums*, Goffman analyses a particular modality of identity construction—through presentation—in 'total institutions'. These are 'symbolized by the barrier to social intercourse with the outside, and to departure that is often built right into the physical plant, such as locked doors, high walls, barbed wire, cliffs, water, forests and moors'.[12] Similarly, when an individual joins an air force, they become subject to various control measures that reflect the unique nature of military service. This experience is particularly evident during basic training, which often requires trainee officers to live on base and closely coordinate their activities. There is an instrumental benefit to developing close-knit and socially cohesive military forces, and 'institutionalising' members through dramaturgical identity construction is an effective method.

This chapter does not challenge this method's efficacy. However, questions arise upon closer examination of the particular military identity elements presented and constructed. Following William Connolly, I argue that identity necessarily involves negation: for individuals to recognise themselves as something, they must differentiate themselves from what they are not.[13] Thus difference, rather than likeness, becomes the controlling factor in the identity-formation process. In the context of the sequential method of identity construction described earlier, it seems reasonable to suggest that the content of the identity presented and subsequently performed is of interest if we wish to ascertain how identity might mould behavioural norms in an organisation. Analysis of the data indicates the existence of an idealised air force identity, the content of which broadly sketches an image of a male, hyper-masculine, classical warrior.

Normalised male

The idea of a normalised male identity emerged in all focus groups, both mixed and single-sex. However, it is fair to say that the all-male groups display the most candour in critiquing the role of women in the US Air Force (USAF). For instance, a male USAFA sophomore cadet stated:

> I think that's just always naturally going to be how males see females, as, you know, um . . . males are put . . . not in charge, but like, have the responsibility of, protecting the females, like the head of the house.[14]

An analysis of the language deployed by some of the males in these focus groups revealed that much of this rhetoric was delivered from a position of ownership (both of the air force and 'our women'), with male cadets often questioning why 'we' should accommodate and change 'our' military to incorporate visible minorities such as women: 'yes they're capable of it [military service], but the question comes down to whether or not we *should* try to integrate them'.[15]

Focus groups in the RAF yielded less-striking examples of a normalisation of the male. However, the theme still emerged in the content analysis of the discussions. A female RAF College Cranwell cadet noted that 'there are males who definitely try to usurp your authority and will just speak straight over the top of you when you're in . . . a leadership position'.[16] Another cadet in the same all-female focus group recalled her experience:

> In fact, when we were on exercise last week . . . and obviously you're in a group of people who don't know you very well, and you don't know them very well . . . and I was one of two girls . . . and I was either given no job, or I was given media and families. I wasn't trusted with nav or comms or anything like that until the last day when one of the lads on our flight, who

obviously knew that I'm capable and I have the ability to do stuff other than just talk to families, he put me on nav. But he was the only one.[17]

One female cadet used the analogy of a 'stencil' to describe a general homogeneity associated with perceptions of the ideal RAF officer:

> For me, the joke was, before I joined, that . . . I swear there's a stencil somewhere in RAF Cranwell that they have to fit through in order to get in, 'cos they're very similar, if that makes sense . . . like, they all look the same . . . there's a kind of persona and image that comes around.[18]

Hyper-masculine

Along with the concept of the normalised male, hyper-masculinity emerged as a particularly valuable trait during basic training. Often, the concept arose in focus groups through a discussion of whether men and women should have different fitness standards. In basic military training, cohesive teams must be built in a relatively short period. This cohesion is commonly achieved by immersing cadets in scenarios that replicate the ground combat environment. On the surface, it seems logical to deploy a military scenario to foster military cohesion. However, analysis of how the cadets interpret the necessary skills and character attributes to succeed in such an activity led to the emergence of some interesting perspectives. For example, a male USAFA sophomore cadet noted:

> We have something called MOUT, where, basically, we just shoot wax bullets at each other down in this mock-up village, and obviously, it's not real combat, but . . . we had two girls on our team who we're [practising] with now, and usually, we sandwich them in the middle of the team. I don't know why; it's not that they're bad at clearing doors or anything; it's just that I have to choose between sending in, you know, 200lbs Jack dude as opposed to this girl . . . who's just as capable. I would rather send in the guy to run in the door first, to, you know, get shot at and say, 'okay, you're gonna absorb all the bullets, and after you get shot, we're gonna come in' . . . it's just human nature.[19]

These comments suggest that, rather than being viewed as an exercise in team-building (which may have prompted a more egalitarian approach to the activity), the emphasis fell more on the tactical benefits of protecting women, in line with 'human nature'.

More broadly, analysis of the data (particularly from the male-only groups) suggests a pervasive suspicion of any modification to the physical standards associated with roles in the military. As a member of an all-male focus group suggested:

> I think you can't adjust [physical standards], just to make someone happy, or to make it so that someone else can join. It's just like saying, 'well then, let's just allow people who don't have arms to be infantrymen'; you can't do the

job, but should we adjust the rules so that this group of people can do some-thing [when] they're not physically able to do the job as well?[20]

The fact that academy cadets, the majority of whom have no experience in the military and are thus broadly ignorant of the practical physical demands of typical air force operations, express such attitudes suggests an absence of context in the generation of the training programme in basic officer training. Put simply, absent of any wider knowledge of the purpose behind such training, some cadets assume this training represents their future experience in the air force, the result of which is a naturalisation of certain ideal identity traits.

Warrior

Unlike cadets at the US Military Academy West Point, relatively few USAFA graduates will find themselves conducting infantry-style assaults on villages, as the exercise described earlier simulates. This fact notwithstanding, the inclusion of traditionally 'military' activities such as minor infantry tactics constructs an idealised model of an air force officer that emphasises the attributes of a classi-cal 'warrior'. This concept is not unique to the Academy. The first stanza of the 'Airman's Creed'—a statement designed to articulate the warrior ethos of the USAF—declares: 'I am an American Airman. I am a Warrior. I have answered my nation's call'.[21] Of the various definitions of the term warrior, focus group responses suggest a more focused understanding of the term among officer train-ees than simply a descriptor for one who participates in war:

> I think that we get a lot of, you know, historical role models; we have, you know, on our walls, Medal of Honor recipients, and some buildings named after specific officers . . . as freshmen here we're drilled on the history of the place, the history of these different war heroes, basically, and I think the underlying idea is that, you know, try and be like them. They don't *say* that, but I think, why else would we study that stuff?[22]

Participants in the all-male RAF focus groups reinforced a conceptualisation of a warrior as one who engages directly with an enemy:

> A key component would be—for an RAF Officer—is to be courageous and a war fighter . . . in fact war fighter is the number one principle of being an RAF Officer . . . and we've [only] shot 42 rounds.[23]

The linkage of the instrumental use of lethal weaponry with 'being a war fighter' was common in focus group dialogue. Discussions relating to female participa-tion in the military invariably used ground combat units such as the RAF Regi-ment (the ground fighting force of the RAF that today provides a range of force protection effects underpinned by its air-minded ground fighting capabilities) rep-resented a type of benchmark for physical fitness standards.[24] Similar comments were made in USAFA focus groups on the relative merits of 'lowering' physical standards to increase the representation of women in the organisation. Physical standards seem to be viewed as sacrosanct, infallible indicators of an individual's utility in combat—the measure of a warrior—rather than an artefact of training regimes that have developed over many years, mostly in the absence of women. The absence of context in the development and delivery of training may contribute to a coalescing of 'war fighter' with the direct application of lethal, kinetic battlespace effects. Interestingly, official RAF discourse is more circumspect:

> Being a war fighter first is important, as it is the core business of the RAF to exploit the air environment for military purposes. . . . There are more than sixty different career specialisations in the RAF used to provide expeditionary air power and every person engaged in those specialisations must contribute to the precise campaign effect at range in time that is required—in other words, they must be a war fighter. To do that well they will have to be highly skilled in their specialisation but focussed on the purpose of the RAF . . . traditional RAF distinctions between those who fight and those who support are breaking down. Not only is the 'support space' vital to the 'battlespace' but so is the 'business space'.[25]

Here, the role of warrior is assigned to all RAF specialisations, including those inhabiting the 'business space'. In this context, to be a war fighter is to carry out your duties whatever they might be, since all contribute to the exploitation of 'the air environment for military purposes'.[26] Rather than representing the *true* 'war fighting' cadre within a larger bureaucracy, the RAF Regiment member merely inhabits a different point on the war fighting continuum to the Personnel Support Officer. In practice, however, ideal identities in basic training seem to run counter to such an understanding. As typified by the experience of the female trainee not assigned instrumental duties during field exercises, for some undertaking officer training, ideal identities look quite different.

When these three ideal attributes are considered within Connolly's conceptualisation of identity, it is plausible to suggest that the normalised male, hypermasculine, warrior persona is being deployed as the template against which assessments are made as to an individual's 'fit' in the organisation. In other words, the oppositional construct is harnessed internally to police and categorise members according to certain characteristics. Popular discourse sometimes suggests that these attitudes are brought to the military by years of socialisation in a heteronormative world that encourages each sex to replicate certain characteristics.[27] That may well be true, but perpetuating such implicit assumptions of value may have the effect of forestalling efforts that work towards a military that attracts and harnesses the widest possible cross-section of Australian society.

Policy implications

Too often, strategists overlook the sociological dimensions of air and space power. Such oversight is understandable, as the manifestations of air and space power's instrumental effectiveness are compelling and often spectacular. The current pau-city of research focusing on Australian military sociology is a testament to this; however, as this research suggests, sociological approaches such as the sequential model of identity presentation and performance presented in this chapter can provide new avenues for exploring the possible causes of under-representation of certain groups in Defence. Is it possible, for instance, that the idealised (and thus presented and performed) Australian identity of a fighter pilot incorporates characteristics that women cannot successfully or credibly perform? Data gathered thus far in the UK and US lend weight to this hypothesis by highlighting the role of identity as a type of gatekeeper within an air force. Indeed, the evidence suggests a double bind that traps those displaying perceptible difference during basic training, as articulated by a female cadet at the USAFA:

> I think they're [female cadets] judged a little bit more harshly in their leadership positions, because [you're], (A): a pushover, or . . . for want of a better word . . . pardon my French . . . a bitch. And so, I think it's very rare that a female has the same leeway as a man might . . . from my experience, the criticism, in general, is a little bit more aggravated if you're a woman.[28]

Another cadet articulated her experience as an African American woman:

> They made fun of me because of cultural differences, and they were just like, 'well, I don't think that you know, women should be in the military because of blah', and I was like, well, that happened, like, thirty years ago. Or they'll be like, 'you know . . . I think the only reason why black people are here is because it's more of a pity thing and not because you guys actually got here on your own merit'. So, they're like, 'yeah, you got the double threat; it's like, you were black, and you were female, and that's how you got in here'. And I was like, no; I worked my butt off to get here. But I mean, you still see it *a lot*.[29]

Thus, faced with a contrasted 'ideal' identity and with an awareness of the consequences of divergence, minority groups may be incentivised to replicate the ideal as best they can to fit in. Thus, senior female cadets reproduce and perform the traditional drill instructor persona at USAFA, berating and shouting in the faces of first years on their induction day.[30] Unfortunately, these performances are often held to a comparatively lower standard or at least viewed differently, as articulated by the cadet who suggested such performances invite labels such as 'bitch'. This issue is not specific to air forces. Still, the consequences are arguably magnified in an organisation whose senior leadership has worked its way to the top through many years of validated and peer-reviewed service.

Recent cultural change initiatives and targeted recruitment strategies provide evidence of a significant effort to build an inclusive and representative force, which is clearly articulated by senior leaders across the services and government.[31] Invariably, the need to 'diversify' is argued based on potential capability gains. This argument is compelling; to compete in an increasingly competitive job marketplace and to be able to deliver the air power effects expected of it reliably, the RAAF must appeal to and harness the talents of the broadest possible spectrum of Australian society. However, the slow rate of change suggests that some stones remain unturned, and identity is one area I believe should be a central focus in enacting meaningful cultural change. A good first step might be to critically examine the identity presented and performed at basic Air Force officer training. Such work should assess its relevance in contemporary air forces where the roles and activities of most members very rarely, if ever, require the specific qualities and attributes of a male, hyper-masculine classical warrior.

Notes

1 The Original Hooters, "The Official Saga: Hooters History," *The Original Hooters*, n.d., https://originalhooters.com/saga.
2 Kenneth Schneyer, "Hooting: Public and Popular Discourse about Sex Discrimination," *University of Michigan Journal of Law Reform* 31, no. 3 (1998): 566.
3 Deborah Lee James, General Mark A. Welsh and Chief Master Sergeant James A. Cody, "Memorandum for All Airmen: Air Force Diversity & Inclusion," 4 March 2015, www.af.mil/Portals/1/documents/SECAF/FINALDiversity_Inclusion_Memo1.pdf.
4 Royal Australian Air Force, "Women in the Air Force," Royal Australian Air Force Website, accessed 7 December 2021, www.airforce.gov.au/our-people/our-culture/women-air-force.
5 See, for example: Scott E. Page, *The Difference: How the Power of Diversity Creates Better Groups, Firms, Schools, and Societies* (Princeton: Princeton University Press, 2007).
6 Australian Institute of Family Studies, *Australian Defence Force Family Census* 1991 (Canberra: Australian Institute of Family Studies, 1991); Royal Australian Air Force, "Women in the Air Force".
7 In 2021, women comprised 51% of all (full-time and part-time) Australian workers. See Australian Government, "Workplace Gender Equality Agency (WGEA) Data Explorer", https://data.wgea.gov.au/industries/1#gender_comp_content
8 Moya Lloyd, *Judith Butler: From Norms to Politics* (Cambridge: Polity Press, 1989): 58.
9 USAFA Male Freshman Cadet, 25 February 2015.
10 Diana Jean Schemo, *Skies to Conquer: A Year inside the Air Force Academy* (New Jersey: John Wiley & Son, 2010), 16–17.
11 Lloyd, *Judith Butler*, 179.
12 Erving Goffman, *Asylums: Essays on the Social Situation of Mental Patients and Other Inmates* (New York: Anchor Books, 1961), 4.
13 See: William Connolly, *Identity/Difference: Democratic Negotiations of Political Paradox* (Minnesota: University of Minnesota Press, 2002).
14 USAFA Male Sophomore Cadet, All-male Focus Group, 24 February 2015.
15 USAFA Male Sophomore Cadet, All-male Focus Group.
16 RAF College Cranwell Female Cadet, All-female Focus Group, 15 September 2015.
17 RAF College Cranwell Female Cadet, All-female Focus Group.
18 Ibid.
19 Ibid.
20 Ibid.
21 T. Michael Moseley, "CSAF Presents Airman's Creed," *Seymour Johnson Air Force Base* Website, 25 April 2007, www.seymourjohnson.af.mil/News/Article-Display/Article/308206/csaf-presents-airmans-creed/.

22 USAFA Male Sophomore Cadet, All-male Focus Group.
23 RAF Cranwell Male Cadet, Mixed Focus Group, 17 September 2016.
24 Peter Jacobs, *The RAF in 100 Objects* (Stroud, UK: The History Press Ltd, 2018).
25 Royal Air Force, "Status of the Leadership Development Program," 2005, PTC/10/1/4/ AMP(943/05) – Annex A.
26 Ibid.
27 Peter Leahy, "These Cadets Are Neither Officers nor Gentlemen," *The Age*, 12 April 2011.
28 USAFA Female Senior Cadet, Mixed Focus Group, 24 February 2015.
29 USAFA Female Senior Cadet, Mixed Focus Group.
30 Tom Roeder, "Cadets Get Fleeting Chance to Smoke Air Force Academy Brass," *Colorado Springs Gazette*, 25 June 2014, https://gazette.com/military/cadets-get-fleeting-chance-to-smoke-air-force-academy-brass/article_cb28ea3e-d9bb-51e5–99d8–13a341155b78.html.
31 See, for example: Department of Defence, *Pathway to Change: Evolving Defence Culture –A Strategy for Cultural Change and Reinforcement* (Canberra: Commonwealth of Australia, 2012).

References

Government publications

Australian Institute of Family Studies. *Australian Defence Force Family Census 1991*. Canberra: Australian Institute of Family Studies, 1991.
Department of Defence. *Pathway to Change: Evolving Defence Culture – A Strategy for Cultural Change and Reinforcement*. Canberra: Commonwealth of Australia, 2012.
Royal Air Force. "Status of the Leadership Development Program." 2005. PTC/10/1/4/ AMP(943/05) – Annex A.

Published sources

Connolly, William. *Identity/Difference: Democratic Negotiations of Political Paradox*. Minnesota: University of Minnesota Press, 2002.
Goffman, Erving. *Asylums: Essays on the Social Situation of Mental Patients and Other Inmates*. New York: Anchor Books, 1961.
Jacobs, Peter. *The RAF in 100 Objects*. Stroud: The History Press Ltd., 2018.
Lloyd, Moya. *Judith Butler: From Norms to Politics*. Cambridge: Polity Press, 1989.
Page, Scott E. *The Difference: How the Power of Diversity Creates Better Groups, Firms, Schools, and Societies*. Princeton: Princeton University Press, 2007.
Schemo, Diana Jean. *Skies to Conquer: A Year inside the Air Force Academy*. Hoboken: John Wiley & Son, 2010.
Schneyer, Kenneth. "Hooting: Public and Popular Discourse about Sex Discrimination." *University of Michigan Journal of Law Reform* 31, no. 3 (1998): 566.

7 The Royal Australian Air Force and the tyranny of training

Tom Frame

When Roger Waters of Pink Floyd wrote the band's most famous song, 'Another Brick in the Wall' (Part 2), it was intended to be a protest song. The lyrics contained the memorable lines: 'We don't need no education; we don't need no thought control'. The target of Water's critique was intellectual conformity, institutional rigidity, and individual oppression. He thought education should have been about freedom, openness, and liberty but found that the schools he attended were their enemies. The purpose of education was preparing students for unquestioning acceptance of their place within a vast impersonal machine, to dutifully perform the functions arbitrarily allocated to them by impersonal forces that assumed an unassailable authority.[1]

In the 40 years since the song was released, much has changed in the objectives and outcomes of formalised schooling. The focus has shifted from teaching to learning and curriculum to capability. In Australia, education has become the chief source of social mobility and economic ascendancy. Few areas of government activity are deemed worthier of a substantial investment than education. This is also true of the Australian Defence Force (ADF), arguably the largest single government sponsor of education and training. The percentage of officers across all three services with a university qualification has increased from less than two per cent in 1967 to nearly 50 per cent in 2017. Across the same period, the percentage of officers with a higher degree has risen from practically nil to nearly 25 per cent.[2] This is a seismic shift in just 50 years and a change that ought to be welcomed.

In the Royal Australian Air Force (RAAF), where the institutional focus has consistently been on harnessing technology, there is a clear preference for training which has come at the expense of education—at least in emphasis—and the confidence that what training delivers in terms of capability does not exist in relation to the benefits of education in terms of capacity. The consequences of a greater emphasis on training are largely imperceptible on a daily basis. Still, they might have led within the RAAF to an instrumentalist approach to problems, thereby depriving the Air Force of the richer intellectual culture that pervades the Army and the ability to articulate the RAAF's remit beyond its own members. It is not a case of 'either/or' when it comes to education and training; it is 'both/and'. Yet, the need to harness technology that has preoccupied the Air Force across the past

DOI: 10.4324/9781003230656-10

60 years seems to have been at the expense of devising a vision for education as the potential basis for a holistic depiction of air power.

This perception is based on research conducted for a volume covering the 50-year relationship between the University of New South Wales (UNSW) and Defence.[3] It is an observation and not a criticism because much can be said in terms of mitigation. As someone whose formative years were spent in the Navy, I would make very similar observations of the Royal Australian Navy (RAN). This chapter aims to draw attention to an element of air power that is often overlooked: the air force's ability to explain its application to the parliament, the press, and the people.

We need to start near the beginning and the evolution of a flawed mindset. When ministerial approval to establish the RAAF College was given on 9 July 1947, its charter stipulated that it would provide tertiary education and professional training for RAAF officers who were on permanent commissions. The initial offering was a four-year course that concentrated on both science and mathematics and RAAF subjects and flying training. It appeared that the educational standard of the College's inaugural entry was largely unsatisfactory, with many cadets struggling to reach the required proficiency, especially in science subjects. This outcome partly reflected the uneven character of secondary school education around the country. To its credit, the RAAF did not consider lowering its standards but actively raised them to ensure its officers were not considered inferior to those of the RAN or the Australian Army. The discussion shifted to achieving the right balance in course content between academic education and technical training.

In September 1956, Air Vice Marshal Frederick Scherger, the member of the Air Board responsible for personnel, ordered a review of the College syllabus. Scherger's personal objective was to see the four-year course reduced to three years. The first two years would be devoted to academic subjects, and the third focused on flying training and operational acquaintance. The Chief of Air Staff, Air Marshal Sir John McCauley, made clear his view that 'the ultimate aim should be the attainment of a Bachelor Degree in Science'.[4] The review committee subsequently recommended in November 1957 that General Duties cadets be educated to Bachelor of Science standard, with the University of Melbourne granting degrees to graduates on the condition that Point Cook staff, facilities, and research programmes met its specifications. The rationale was plain. In 20 to 30 years: 'when the first graduates of this new scheme would be reaching high rank, the Air Force will probably be primarily a "missile" Service'. Its leadership 'should have a very broad and advanced education' based on a Science degree 'in order to establish a widely recognised standard which will encourage boys with the high academic aptitude and education which is desirable'.[5] It would be a broad education, but it would be a science degree.

With the goodwill of the University of Melbourne, the College was reconstituted at the beginning of 1961 as the RAAF Academy and was formally affiliated with the University. The affiliation statute explained that 'the Academy shall be an educational establishment affiliated with the University of Melbourne for the purpose of instruction of candidates for the various degrees of the Faculty of

Science and the degree of Doctor of Philosophy of the University'.[6] The course for General Duties officers (the non-pilot cohort) was four years duration. The first three years were essentially a Bachelor of Science degree with a major in physics. There was no choice in subjects studied, with classes spread across four 'divisions': Military Studies, Pure Science, Aeronautical Science, and Humanities. The fourth year of study was designed 'to build a bridge between the pure studies and their application to the Air Force'.[7]

The first few years of the affiliation were challenging for both the academy and the university, with a higher-than-expected failure rate and unexpected requests from cadets to withdraw from the degree programme. The combined attrition rate would later rise to 60 per cent.[8] The university was also proposing an extension of the science programme to four years (with the inclusion of additional pure science and non-science subjects) and wanted more control over the subjects it was being asked to recognise by the Air Force. By July 1964, there was deep disquiet within the Air Board. It had received a paper concluding that 'the aim of the Academy is not being attained'.[9]

> The motivation of cadets to the degree studies suffers because they do not see in the course an obvious application to Air Force requirements. As a result, their performance on the degree studies falls below what they are capable of. There is a tendency among cadets who have completed the degree component part of the course to feel that the training which they have undergone will not be utilised within the RAAF. As a result, they look around for civilian careers, which will make more use of their training.[10]

The lack of a vision for education led to a deficit of application to education.

The purpose of education is not to produce an educated person but to provide a platform for professional Service-oriented training. There is no sense in which education was viewed as acquiring the discipline of critical thought, broadening of the intellectual horizon, the development of cognitive imagination, or the challenging of personal prejudices. Education was seen in almost purely vocational terms. The emphasis was on what a person could do rather than who they would become.

The Minister for Air and former wartime Fleet Air Arm pilot, Peter Howson, wrote to the Vice-Chancellor of the University of Melbourne, Professor Sir David Derham, proposing the development of an applied science degree that would more closely meet Air Force needs, that is, a degree that would be even more narrowly focused. A lack of clarity continued until the end of the 1960s, when the Martin Committee examined the prospect of a tri-Service Academy and described the academic burdens imposed on RAAF Academy cadets by the University of Melbourne as 'unrealistic'. A single degree course was still being offered—the Bachelor of Science—and there were only two majors: mathematics or physics. The Committee explained:

> In our opinion, the RAAF Academy illustrates the difficulties which arise from a close association between a university and a Service college.

A baccalaureate course in Science . . . with its emphasis on physics, is not wholly suited to the needs of the RAAF.

But what were these needs?[11]

Three years later, the long-serving academy warden, Mr Walter Hardy, produced a confidential report entitled 'Decline in Academic Standards, RAAF Academy', which noted that cadets could not see the value of their degree beyond it being a graduation requirement.[12] Nor did the Air Force seem very interested in its content. The '51 per cent syndrome'—the idea that it was necessary only to pass rather than to excel—had taken hold even among cadets capable of achieving a high distinction average in an Honours programme. Notable here is student disinterest or indifference to the personal and professional importance of what Sir Arthur Tange described, when Secretary of the Department of Defence, as a 'balanced and liberal education'; something that would give young officers an ability to be self-learners and a desire to be life-long learners.[13] Consistent with corporate attitudes in the Navy, the Air Force did not promote higher degree study among its younger officers. The prevailing view was that the initial entry education and training pipeline was already too long.

There were also long-standing cultural conflicts. The Academy's historian, Air Vice-Marshal Roy Frost, concluded:

> The problems of amalgamating the spirit of individual inquiry and intellectual integrity with military discipline and unqualified regard for authority were in evidence throughout the life of the Academy. The conflicts between the objective of education which seeks to stimulate discussion and independent thought, on the one hand, and training on the other, which looks to vocational relevance and group orientation, were never satisfactorily resolved from an Air Force point of view.[14]

Given that many non-university graduates had achieved senior rank in the Air Force, Frost thought there were 'reasonable grounds for querying the wisdom of establishing and maintaining' either the RAAF College or the RAAF Academy. Ultimately, the Air Force might have met its needs just as well without the cost of an educational institution to provide a university degree. But perhaps, he mused, many of those with degrees who left the Air Force might have done so before the vocational value of their education was apparent.[15]

Frost's observations suggest the RAAF did not have a coherent view of education and how and why training was different. While there was a general endorsement of the need for education (it was unlikely that any officer would run the risk of being called a philistine by arguing against it), there was nothing that resembled a consistent approach to education (especially in the humanities). The Air Force lacked the ability to articulate its contribution to the national interest, to the nation's strategic needs in a time of contested outlooks, or to public perceptions of what the RAAF needed in terms of technology and material support. One could argue that these tasks fall to senior officers, but the ability to think, articulate, and persuade is refined over decades. Every officer is confronted by the need to give an account of their Service and its remit throughout their career.

This chapter does not canvass changing views within the RAAF over the last three decades because I have not encountered a substantive statement on education since 1980 that would dislodge my general impression of the place of education in the RAAF: it remains strongly vocational and is largely influenced by technology management. The RAAF has produced very few public intellectuals of the calibre of Professors Robert O'Neill and David Horner (both former infantry officers) or Admirals James Goldrick and Peter Jones, only a handful of officers write for publication, and attracting RAAF personnel to conferences is a perennial challenge. It is possible to argue that the RAAF is busier than the Navy or the Army, or that the nature of Air Force service is different. In my view, institutions invest in what they value, and it is difficult to argue the Air Force has been interested in developing an intellectual culture to rival that of the Army and, to a lesser extent, the Navy.

The Army has traditionally placed greater importance on producing officers who can think expansively and reason creatively to influence government decision-making in its favour, to persuade the public of its intrinsic worth, and to manage internal disagreement about how force can be effectively and efficiently applied to secure national interests. The Army's establishment of the well-resourced Army History Unit, the Chief of Army's history conference, the Centre for Army Lessons, and the Army Research Centre testify to the army's willingness to invest in assessing past performance and participating in the contest of ideas. The Navy's Sea Power Centre and the annual King-Hall history conference are reflections of the Navy's determination to not be left behind in the quest to 'sell' the Service's story to the parliament, the press, and the people, and to ensure it harvests the limitless potential of the human mind.

Substantiating my claim of over-emphasis, in a speech delivered by then Chief of Air Force to the Australian Strategic Policy Institute (ASPI) in July 2016, Air Marshal Leo Davies outlined the RAAF's ten-year strategy. Davies explained that it rested on five distinct vectors. Two of these vectors—people and communications—were described as technology-enabled and training-supported. There was no mention of education because the main thrust was the technology-training axis. Technology provides the strategic edge; training ensures its effective application. He does not discount the value of education, but training is pre-eminent.[16]

Today, the RAAF appears to be unsure about the connections between education and employment. How does education prepare a RAAF member for the requirements of their Service? It has been said that training is a necessity and education is a luxury; the former can be mapped alongside duty statements while the latter is indirectly applicable and, only then, in a very generalised way. In the context of imprecise outcomes, it is prudent to maximise training and minimise education. Conversely, it has been argued that training relates to vocation while education serves to build character; as officers and non-commissioned officers must lead, and leadership draws directly on character, the need for education is practically limitless. As training relates to the performance of specific tasks and education focuses on understanding the 'big picture', assessing possibilities, and exploring potentialities, an uneducated officer class will lack initiative and insight.

Although the character of ADF service has changed substantially over the past 50 years and the demands on ADF officers have evolved considerably, the nature of leadership and the expectations of command have remained constant. While it could be argued that the need to promote education may have changed because its importance is acknowledged in all sectors of the Australian community, the counter-argument is that the potential of education to transform people and organisations has not changed and the Air Force has yet to exploit endless possibilities.

The principal tensions are the distance between cause and effect—an education takes time to acquire, and its tangible benefits are not immediately apparent—and discerning where and how a well-educated workforce will enhance policy development and operational outcomes. Given that most organisations measure what they can measure rather than attempt to measure what matters, long-term investment in education almost invariably suffers from the imperatives of short-term expediency. I could conclude this chapter by pleading for closer attention to the many individual and institutional advantages of a wide-ranging education in the liberal democratic tradition. It is a well-rehearsed petition that often sounds like self-interest when promoted by academics. My preference is to finish with an observation that will strike a chord with those more likely to be persuaded by a consequentialist argument.

Since the office of Chief of the Defence Force Staff (later renamed Chief of the Defence Force) was established in 1976, there have been fourteen incumbents. Just three have been Air Force officers: Air Chief Marshals Sir Neville McNamara, Sir Angus Houston, and Mark Binskin. By comparison, the Army has provided eight officers. Furthermore, former Army officers have been over-represented in Federal and state parliaments and a range of senior official posts. It is certainly true that Air Force officers have succeeded in securing a greater proportion of Commonwealth funding for acquisitions and operations for their Service. Still, Army officers have been substantially more influential within government circles and their legacies have been more lasting. The challenge remains for Air Force to match technological supremacy with greater influence in shaping of national policy. Education is key to that outcome.

Notes

1 Roger Waters, *Another Brick in the Wall (Part 2)* (London and New York: Harvest/ Columbia, 1979).

2 The long-term trends in primary and higher degree completions are discussed in Tom Frame, *Widening Minds: The University of New South Wales and the Education of Australia's Defence Leaders* (Sydney: UNSW Press, 2017), 436–37.

3 Frame, *Widening Minds*.

4 Quoted in R. E. Frost, *RAAF College and Academy, 1947–86* (Canberra: Royal Australian Air Force, 1991), 35.

5 A copy of this report was included in a folder compiled by Mr Len Hume marked 'Chronology of Considerations and Events Leading to the Establishment of the Australian Defence Force Academy', Academy Library Special Collections 85/327(1).

6 Quoted in Frost, *RAAF College and Academy*, 101. Frost does not provide details of the documents upon which he relied for his history.
7 Quoted in Frost, *RAAF College and Academy,* 50.
8 Frost, *RAAF College and Academy*, 53.
9 Ibid, 52.
10 Ibid, 52.
11 Quoted in Frost, *RAAF College and Academy*, p. 59.
12 Frost, *RAAF College and Academy*, 59.
13 Frame, *Widening Minds*, 137–38.
14 Frost, *RAAF College and Academy*, 101.
15 Ibid.
16 Leo Davies, "Air Force's Strategic Plan for the Next Decade" (Speech, National Security Dinner, Australian Strategic Policy Institute, Canberra, 19 July 2016), www.aspi.org.au/video/national-security-dinner-2016.

References

Published Sources

Frame, Tom. *Widening Minds: The University of New South Wales and the education of Australia's defence leaders*. Sydney: UNSW Press, 2017.
Frost, R. E. *RAAF College and Academy, 1947–86*. Canberra: Royal Australian Air Force, 1991.
Waters, Roger. *Another Brick in the Wall (Part 2)*. London and New York: Harvest/Columbia, 1979.

Part Three
Technology and Air Power

8 Human, Organisational, and technological lessons from air power and joint operations in major conflict

Charles Vandepeer

The Royal Australian Air Force (RAAF) is currently in the process of the most significant technological change in its history as it transforms into a fifth-generation air force; that is, a fully networked force that 'exploits the advantages of an available, integrated and shared battlespace picture to deliver lethal and non-lethal air power', and provides the strength necessary to combat the 'increasingly complex' threats posed by warfare in the Information Age.[1] To meet these requirements, the RAAF initiated Plan Jericho, a project that aims to confront the challenges posed by new and emerging capabilities and technologies within a changing strategic environment.[2] To deal with this broad scope of challenges, Plan Jericho outlines the requirement for the RAAF to 'develop an innovative and empowered workforce', and 'embrace innovative thinking and prepare to harness the potential of emerging technology rapidly'.[3] Concerns regarding a more contested military environment in which potential adversaries share similar technological capabilities provide an impetus for the RAAF to innovate and improve.

However, the RAAF has a gap in its experience, as its more recent operational experience differs from the type of future contested air domain anticipated by Plan Jericho. Recent operations have seen the Australian Defence Force (ADF) and its Coalition partners fight against lethal but technologically inferior non-state actors who have countered their disadvantages using low-tech but deadly tactics and capabilities. Consequently, recent and current operations do not necessarily reflect the contested air environments envisaged under Plan Jericho. Indeed, it has been decades since the RAAF embarked on operations without control of the air, often because they operated as part of US-led coalition forces. Becoming overly focused on lessons from recent operations risks missing the opportunity to learn lessons from conflicts where Allied forces faced technologically comparable and, at times, technologically superior, state-based adversaries.

While there have been recent developments in technology, platforms, and capabilities, significant technological and capability transformation in warfare is not unprecedented. The Second World War witnessed arguably the most significant period of technological and capability development and the military deployment of new and adapted capabilities in human history. Radar, radio, jet engines, computers, signals intelligence, massed long-range heavy bombers, and dedicated special force units were some technologies and tactics that transformed warfare

DOI: 10.4324/9781003230656-12

and required new and novel approaches to both exploit and counter such advancements. This chapter considers several case studies in air power and joint operations where military personnel and operations scientists adapted and employed existing and new technologies, as well as new tactics and operational methods, to achieve decision superiority over formidable state adversaries. Although this chapter does not allow more than a brief insight, it highlights some pertinent and important lessons.

RAF Fighter Command: Success Through Foresight and Fear

In 1932, British Parliamentarian (and later Prime Minister) Stanley Baldwin uttered the phrase 'the bomber will always get through' in discussing the significant challenges posed by the threat of aerial bombing.[4] This phrase, building on the air power theories of Giulio Douhet, clearly had a significant impact. In 1936, the Royal Air Force (RAF) established Fighter Command, led by Air Marshal Hugh Dowding, to address the threat presented by bombers and the significant risk they posed to Great Britain's defence. When Fighter Command was established, neither the Hurricane nor Spitfire had yet come into service, and only early-stage consideration of the potential use of radar to locate aircraft had been made. Harold Lardner, one of the operations research scientists involved in establishing what was to become the world's first integrated air defence system (IADS) and ground-controlled intercept capability, emphasised that Luftwaffe bombers confronted Britain with 'a technical problem for which they could see no solution':

> Basically, the problem stemmed from the fact that no part of the British Isles lies further than 70 miles from the coast—a scant 17 minutes of flying time for the German bombers that were later to be used . . . there was no means known at that time for providing the warning needed for defence against enemy air attack (i.e., warning and tracking of approaching bombers) that could extend sufficiently far beyond the coastline so that defending fighter aircraft would have time to take off, gain altitude, and engage the enemy before he could penetrate to most of the vulnerable centres.[5]

Radar offered such a possibility, and through the initiative and support of only a handful of senior civilian and RAF leaders (including Dowding), resources, personnel, and funds were made available to provide a solution to the problem. Work started in December 1934, when the Committee for the Scientific Survey of Air Defence headed by Sir Henry Tizard was established. Over the next few years, the potential of radar was identified and then developed. By September 1936, the first basic air exercise in connection with radiolocation was run.[6] Fighter Command, which recognised the risk that German bombers presented, proactively welcomed, and worked with scientists. This collaboration led to the establishment of Operational Research, a new scientific field of research that emerged out of the specific military requirement to defend Britain against air attack and efforts to utilise the emerging radar technology to address this problem.[7]

The RAF and operations scientists were dealing with entirely new technologies, coupled with new platforms (Hurricanes and Spitfires as they came online) and training people in new skills (radiolocation tracking). The development of ground-controlled intercepts, in which the ground stations provided RAF fighter aircraft with location guidance to locate and engage incoming Luftwaffe bombers, was a matter of trial, error, learning, and improvement. There was no complete IADS package that came ready-to-use and ready-to-apply. Instead, the combination of new technologies, platforms and skills required an iterative learning approach to maximise the capabilities of a system being designed, adapted, and improved. The desperation in and motivation behind the testing and development of these technologies, processes, and tactics is clear in the following description of efforts during 1937, when

> the development of the fighter tactics for use with radar had begun, and the fighter crews were trained in their use before radar itself was developed. If the RAF had waited to work out fighter tactics until radar had been developed to a considerable degree, there would not have been time to elaborate the new tactics and train the crews. Without the far-sighted development of tactics at Biggin Hill, the RAF could not have efficiently utilised the advantage given to them by radar in the Battle of Britain.[8]

This observation is profound in its implications: the RAF actively developed and honed flying tactics before radar technology was available. This level of proactivity had strategic and clear outcomes. By the time the Battle of Britain began, Fighter Command had used the very short time available since its establishment to its best advantage.

Desperation appears to play a significant part in such creative solutions. Gary Klein argues that one way to gain insight is through creative desperation, as the situation dictates that something different must be tried.[9] Instead of taking a 'failure is not an option' approach, recognising the real potential for failure appears to encourage creative problem-solving. The culture of desperation and genuine openness was reflected in another aspect of the interaction between the military and civilian personnel who worked together during the initial German raids over Britain. It was not a matter of simply delivering technology and leaving it to the operators to figure out; operations researchers were actively involved alongside military personnel in operations rooms during the earliest Luftwaffe raids on Britain. These new technologies and systems needed to be used by people, and these elements needed to be effectively combined in operations to achieve an advantage over the adversary. Early raids allowed the RAF and operations scientists to learn from their successes and failures and incorporate these lessons into actual performance.

Consequently, the high levels of efficiency and successes achieved by radar stations during the Battle of Britain were directly attributable to operations scientists who analysed almost every failure to intercept daylight Luftwaffe raids from October 1939 and fed these lessons back into radar operations to improve performance.[10] The development of the first IADS and ground-controlled intercept resulted from foresight and fear. We can contrast Fighter Command's development of tactics with Bomber Command's early years, particularly the difference in attitudes towards scientific support and objective performance measures perhaps best underlies the differences in culture.

Bomber Command: Theory Meets Reality

If Fighter Command's fear spurred them into action and seeking solutions, Bomber Command had a different perspective, reflected in its early years. Britain's chief scientist for Intelligence in the Air Ministry, Reginald Victor Jones, observed that Bomber Command basked in the doctrine that the bomber would always get through and saw little need for scientific support in the early years of the war. Contrasting the experience operations research scientists had at Fighter and Bomber Command until late 1941, Jones noted:

> [A]t Fighter Command we were immediately welcome, and the Commander-in-Chief would readily see us; at Bomber Command it was more like visiting a gentlemen's country club—we would be courteously heard and entertained but would leave with the impression that what we said would have little effect.[11]

Indeed, although Bomber Command had not sought scientific support, they had been offered such assistance. In September 1939, the Director of Scientific Research sent an officer to Bomber Command but recalled them within months due to insufficient demand for their services.[12] At the same time that Fighter Command had scientists stationed within operations rooms, actively contributing to the development of radar, fighter tactics and ground-controlled intercepts, Bomber Command could find no use for such scientists. It was not as if Bomber Command did not have significant problems or challenges in achieving their mission at the time, principal among these issues was the ability to navigate, locate and prosecute their targets accurately.

In late 1941, however, things began to change. There were two important reasons for this change: increasing numbers of bombers were being shot down with the loss of aircrew and aircraft, and new aerial photography capabilities showed that bombers were rarely finding their targets.[13] However, aerial photography results were not initially welcomed by Bomber Command leadership. As Taylor Downing observes:

> For many months, photographic reconnaissance of sites that had been bombed continued to show that the damage caused was negligible. This was not what the chiefs of Bomber Command wanted to hear, and they did not believe it could be true. They came up with a variety of reasons to disprove

the photographic evidence. The photographs were of too small a scale to be able to spot the damage. The photo interpreters did not know what they were looking for. Some damage assessments came back with a note in the margin saying simply: 'I do not accept this report'.[14]

Despite the technological advantage of aerial photography capabilities, the unwelcome news such analysis produced initially prompted its disregard by Bomber Command because it did not fit the desired reality. However, photography reconnaissance enabled the closure of the feedback loop that aerial bombing required, namely the ability to evaluate performance objectively. Using photographic reconnaissance, the Butt Report in August 1941 concluded that a series of raids on the Ruhr resulted in only one-tenth of the RAF bombers finding their way within five miles of their targets.[15] One estimate suggested that 90 per cent of bombs dropped between 1940 and 1941 (some 44,737 tons) probably missed their targets entirely, and these poor results came at the cost of RAF aircrew's lives, loss of aircraft and wasted munitions and effort.[16] Ultimately, due to the weight of evidence supporting the poor performance of strategic bombing coming from outside their Command, Bomber Command eventually became open to scientific advice. Jones found that operations scientists had similarly strong relationships with both commands by the end of the war.[17]

Paul Kennedy argues that aerial photographic reconnaissance, rather than the more inconsistent Enigma intercepts, provided the key information source for evaluating and improving aircraft bombing 'because it was consistent, technical, and objective'.[18] The improvement of this aerial capability provided the key to improving the application of aerial bombing. However, this was possible only once Bomber Command reluctantly accepted the results, and only in conjunction with additional improvements, technologies, and organisational strategies. People were also central in the design, implementation, and adaptation of these developments. The importance of the combination of technologies and people in achieving outcomes was emphasised in *Science at War*, a 1947 review of the impact of scientific thinking in the British war effort. Tellingly taking the example of aerial bombing of targets, the government publication observed:

> Organisation and the mode of tying things together is often more important than improvements in individual weapons. Technical excellence may be wasted by strategical nonsense. For instance, very accurate bombing is no use if you bomb the wrong targets, through failure to choose or find the right ones.[19]

The ideas that the bomber would always get through and that success was inevitable appear to have significantly influenced thinking and mindsets within Bomber Command. The result was the waste of lives, platforms, and weapons when Britain could least afford it. Perhaps most disturbingly, Britain's lack of accurate navigation systems for strategic aerial bombing had been identified as early as the First World War but entirely ignored during the interwar period.[20] Indeed, much evidence and data pointed to the requirement for accurate navigational aids. As Randall Wakelam noted, between 1937 and 1939, there were 478 instances of aircraft being forced down because the pilots had become lost within the United Kingdom.[21] Still, it was not until 1938 that an air navigation office was established within the Air Ministry. One observer noted that 'the Air Staff simply did not appreciate the need for accuracy in navigation, but merely assumed that their bomber crews had such an ability.'[22] There had been efforts to raise awareness of navigation problems. However, these attempts were ignored by Bomber Command, which assumed that their aircrew had the necessary skills to navigate to targets accurately.

Operation Overlord: Failure as a Genuine Possibility

Operation Overlord, the Allied plan to conduct a contested amphibious landing in occupied France, required multiple nations' naval, land, and air forces to operate as a coordinated joint force on a scale and scope never achieved in human history. The size of the Allied invasion force assembled in England, which included some 4,000 ships, 11,000 planes, and nearly three million armed forces personnel, eclipses any operation since.[23] The history of amphibious landings against prepared and well-armed adversaries had been one of very mixed success. Disasters such as the Dieppe raid in 1942, which resulted in significant loss of life and equipment (for the Allied armies, navies and air forces involved), made it undeniably obvious that the failure of Operation Overlord was a realistic possibility. Until the D-Day landings (and for some time after), the possibility of failure was a genuine concern. If there was any doubt as to then Major General Dwight Eisenhower's recognition of the potential for failure of Operation Overlord, the message he penned on the eve of the landings (which was to be released in the event that the landings were unsuccessful) reveals a man deeply conscious of the possibility of operational defeat.[24]

It appears that genuine recognition of the real potential for operational failure, similar to that demonstrated by Fighter Command, spurred efforts to promote innovation and deliberate and dedicated efforts to overcome likely challenges. The inability of Allied tanks and armoured cars to get off the beaches of Dieppe led to the development of a diverse range of purpose-built tanks and armoured vehicles that could operate on the sand, breach enemy defences, and overcome obstacles. These vehicles, dubbed Hobart's Funnies, are credited with helping Allied personnel get off the beaches of Normandy. General Sir Percy Hobart, who oversaw the development of these vehicles, was recognised with having the genius of being open to ideas and a willingness to listen to anyone; 'if you had a good idea, he'd listen to it', irrespective of whether it came from civilian or military personnel, and regardless of their rank or status.[25] Perhaps this openness stemmed from his own experience of being forcibly retired as a star-ranking officer early in the Second World War (reportedly because of his unconventional ideas on armoured warfare), only to enlist as a lance corporal in his village's defence volunteer unit and soon after be reinstated to star rank at Prime Minister

Winston Churchill's direct intervention.[26] Interestingly, it appears that the prime minister only became aware of Hobart's situation when Captain Liddell Hart published a newspaper article under the banner 'We've Wasted Brains'.[27] The fact that Hobart's Funnies came to play such an important role in the Normandy landings reflected the brilliance behind their design and their ability to be effectively integrated with existing technologies, platforms, and people to help achieve overall operation success.

In recognition of the enormous difficulty of the task at hand, and the requirement to develop a robust and successful plan, the need to work with experts—at whatever rank they were to be found—was apparent. The ability to engage experts was reflected in the choice of headquarters staff and planners and those experts in weapons and tactics who would be crucial to the success of Operation Overlord. Major James Goodson, an American citizen who had flown in the RAF and then in United States Army Air Force (USAAF), was a recognised Fighter Ace. In addition to 15 confirmed enemy aircraft kills in aerial combat, he earned the nickname 'King of the Strafers' for successfully destroying 15 Luftwaffe aircraft on the ground during attacks on enemy airfields. Goodson provided the following insight into a meeting he had been called to attend alongside other officers, including fellow Majors, Captains, and a Lieutenant:

> We had a meeting at Debden with Eisenhower . . . and all the brass to discuss our role on D-Day. It was an eye-opener to me, I realised why Ike had been chosen to be supreme commander over more senior generals. It was a lesson in leadership and motivation. He went around the table and asked for everyone's input. No officer was too junior; no comments were too inappropriate not to be listened to. Only once did Eisenhower cut anyone off. When someone said: 'Ike, I've got a great idea', he replied, 'It's too late for great ideas. We now have to make sure the plans we have work'.[28]

Eisenhower's willingness to have his plan questioned, challenged, and critiqued by mid-ranking and junior officers is insightful. The risk of failure was enormous, so it appeared far better for the plan to be critiqued by those who would be fighting rather than waiting to give the enemy the final say.

Paul Kennedy's consideration of Allied successes across numerous operational challenges in the Second World War supports the importance of a senior leader's ability to draw out the expertise, ideas, and suggestions from across rank levels. Kennedy makes the following insightful observation:

> In sum, the winning of great wars always requires superior organisation, and that in turn requires people who can run those organisations, not in a blinkered way but most competently and in a fashion that will allow outsiders to feed fresh ideas into the pursuit of victory. None of this can be done by the chiefs alone, however great their genius, however massive their energy. There has to be a support system, a culture of encouragement, efficient feedback loops, a capacity to learn from setbacks, an ability to get things done.

And all this must be done in a fashion that is better than the enemy's. That is how wars are won.[29]

Even in an environment of enormous technological advancements, the human aspects of listening, learning, encouraging, and empowering people across rank levels appeared critical to the Allies military success in the Second World War.

Lessons

A brief consideration of the application of air power and joint operations in conflicts fought against technologically similar adversaries provides some key insights for the RAAF and the ADF. Recognition of the risk of failure appears to be a good driver of insights, innovations, and a willingness to listen to outside perspectives, whereas overconfidence seems to lessen openness to internal or external critique. This chapter's brief consideration of Fighter Command and Operation Overlord reinforces the argument that a genuine acknowledgement of the potential for failure drives problem-solving and innovation. In contrast, overconfidence in one's abilities and performance can lead to a blindness to actual failure, even in the face of evidence of shortcomings. Indeed, following the surrender of French forces in 1940, it was observed that '[a] naïve belief in invincibility might have some value in morale, but, as experience in France has shown, it is a dangerous guide in strategy.'[30]

Another lesson that may be learnt from these case studies is that environments for ideas and innovation do not just happen; leaders must actively develop and encourage suggestions and critique in the way that Dowding, Hobart, and Eisenhower actively or openly welcomed feedback and suggestions. Scientists had to be welcomed, ideas discussed, and positive environments fostered. People need to be genuinely listened to, and leaders must play a deliberate role in these processes. Talent can be overlooked or wasted if such processes are not supported, as happened with the initial assistance of operations science at Bomber Command or Percy Hobart's removal from service.

In the examples discussed here, as with the observations from numerous Allied successes during the Second World War, solutions were cumulative and iterative. Successful operations against a technologically capable state-adversary resulted from a combination of factors among which people were central. People, both military and civilian, were responsible for applying and adapting existing or new technologies to a defined problem more effectively than the adversary. The importance of a *combination* of technologies, ideas, and tactics in defeating a comparable adversary suggests a warning against the idea of 'magic bullet' solutions. The iterative nature of developments in air power and joint operations also highlights that technologies and processes do not come with pre-ordained solutions to evolving challenges, which leads to the idea of out-learning adversaries.

Ultimately, conflict is a battle of opposing forces. Writing just two years after the end of the Second World War, Crowther and Whiddington argued that 'one reason why Hitler failed is that he was out-of-date'. Whereas the Allies developed

Operations Research and brought scientific analysis to the performance of operations, Hitler held onto a more romantic view of conflict, supported by myth and emotion over facts.[31] The Allied approach proved more effective because leaders, commanders, and operators were realistic in appreciating failure, which indicates the importance of learning. That is, an ability to *out-learn* an adversary in the application of air power and conduct of joint operations. However, this ability to out-learn an adversary is based on realistically understanding the situation, acknowledging failures, and recognising the genuine potential for failure.

Conclusion

The establishment of the world's first IADS through collaboration between Fighter Command and scientists demonstrates that defence forces can be proactive; they can learn and develop capabilities without needing first to lose lives, platforms, or battles. Although this is an incredibly positive observation that is directly relevant to Plan Jericho, success does require a genuine recognition of the real potential for failure, the proactive actions by decision-makers, the importance of outside expertise, and the requirement for developing environments in which ideas and suggestions are actively encouraged from all rank levels. None of these lessons is necessarily easy to implement and might not even guarantee success against a formidable adversary. Nevertheless, a recognition that failure is a genuine option appears to offer at least the potential for genuine innovation and the opportunity to out-learn future adversaries, which history suggests might just provide the basis for victory over technologically comparable adversaries.

Notes

1 Royal Australian Air Force, "Fifth-generation Air Force," *Air Force*, www.airforce.gov.au/our-mission/fifth-generation-air-force.
2 Royal Australian Air Force, *Plan Jericho* (Canberra: Air Power Development Centre, 2015), 2.
3 Royal Australian Air Force, *Plan Jericho – Program of Work 2016: Transforming Air Force's Combat Capability* (Canberra: Royal Australian Air Force, September 2016), 1, 7.
4 Quoted in Brett Holman, "The Air Panic of 1935: British Press Opinion between Disarmament and Rearmament," *Journal of Contemporary History* 46, no. 2 (2011): 292.
5 Harold Lardner, "The Origin of Operational Research," *Operations Research* 32, no. 2 (1984): 468.
6 James Crowther and Richard Whiddington, *Science at War* (London: HMSO, 1947), 6.
7 Lardner, "The Origin of Operational Research," 466.
8 Crowther and Whiddington, *Science at War*, 10.
9 Gary Klein, "Leverage: How We Spot Opportunities," *Psychology Today* [blog], 27 June 2014, www.psychologytoday.com/au/blog/seeing-what-others-dont/201406/leverage.
10 Air Ministry, *Air Publication 3368: The Origins and Development of Operational Research in Royal Air Force* (London: HMSO, 1963), 12.
11 Reginald Jones, *Reflections on Intelligence* (London: Mandarin, 1990), 194–95.
12 Ibid, 196.
13 Ibid, 194–95.

14 Taylor Downing, *Churchill's War Lab: Code-Breakers, Scientists, and the Mavericks Churchill Led to Victory* (London: Little, Brown, 2010), 241.
15 Paul Kennedy, *Engineers of Victory: The Problem Solvers Who Turned the Tide in the Second World War* (New York: Random House, 2013), 103.
16 Crowther and Whiddington, *Science at War*, 50.
17 Jones, *Reflections on Intelligence*, 194–95.
18 Kennedy, *Engineers of Victory*, 138.
19 Crowther and Whiddington, *Science at War*, 117.
20 R. V. Jones, "A Concurrence in Learning and Arms Author," *Journal of the Operational Research Society* 33, no. 9 (September 1982): 781.
21 Randall Wakelam, *The Science of Bombing: Operational Research in RAF Bomber Command* (Toronto: University of Toronto Press, 2009), 15.
22 Quoted in Wakelam, *The Science of Bombing*, 16.
23 Figures quoted at "Message Drafted by General Eisenhower in Case the D-Day Invasion Failed and Photographs Taken on D-Day," *National Archives*, August 2016, https://web.archive.org/web/20160729070911/www.archives.gov/education/lessons/d-day-message/.
24 See: Message, General Dwight Eisenhower, "In Case of Failure," *Docsteach*, 5 June, 1944, accessed 20 April 2022, www.docsteach.org/documents/document/in-case-of-failure.
25 Quoted in Stephen Dowling, "The Strange Tanks That Helped win D-Day," *BBC Website*, 7 June 2016, www.bbc.com/future/article/20160603-the-strange-tanks-that-helped-win-d-day.
26 Prime Minister Churchill's written rebuke to the War Office for overlooking Hobart's talents is worth highlighting: 'I am not at all impressed by the prejudice against him in certain quarters. Such prejudices attach frequently to persons of strong personality and original view. . . . We are now at war, fighting for our lives, and we cannot afford to confine Army appointments to officers who have excited no hostile comment in their careers'. Quoted in Kennedy, *Engineers of Victory*, 269.
27 Basil Liddell-Hart, "We've Wasted Brains," *Sunday Pictorial*, 11 August 1940.
28 James Goodson, *Tumult in the Clouds* (London: Penguin Books, 2003), 152–53.
29 Kennedy, *Engineers of Victory*, 372.
30 Quoted in Crowther and Whiddington, *Science at War*, 33.
31 Crowther and Whiddington, *Science at War*, 119–20.

References

Government Sources

Air Ministry. *Air Publication 3368: The Origins and Development of Operational Research in the Royal Air Force*. London: HMSO, 1963.
Royal Australian Air Force. *Plan Jericho*. Canberra: Air Power Development Centre, 2015. https://airpower.airforce.gov.au/sites/default/files/2021-03/AF14-Plan-Jericho.pdf.
Royal Australian Air Force. *Plan Jericho – Program of Work 2016: Transforming Air Force's Combat Capability*. Canberra: Royal Australian Air Force, September 2016. https://airpower.airforce.gov.au/sites/default/files/2021-03/AF18-Plan-Jericho_Program-of-Works-2ndEd.pdf.

Published Sources

Crowther, James, and Richard Whiddington. *Science at War*. London: HMSO, 1947.
Downing, Taylor. *Churchill's War Lab: Code-Breakers, Scientists, and the Mavericks Churchill Led to Victory*. London: Little, Brown, 2010.

Goodson, James. *Tumult in the Clouds*. London: Penguin Books, 2003.

Holman, Brett. "The Air Panic of 1935: British Press Opinion between Disarmament and Rearmament." *Journal of Contemporary History* 46, no. 2 (2011): 288–307.

Jones, R. V. "A Concurrence in Learning and Arms." *Journal of the Operational Research Society* 33, no. 9 (September 1982): 779–91.

Jones, Reginald. *Reflections on Intelligence*. London: Mandarin, 1990.

Kennedy, Paul. *Engineers of Victory: The Problem Solvers Who Turned the Tide in the Second World War*. New York: Random House, 2013.

Lardner, Harold. "The Origin of Operational Research." *Operations Research* 32, no. 2 (1984): 465–75.

Wakelam, Randall. *The Science of Bombing: Operational Research in RAF Bomber Command*. Toronto: University of Toronto Press, 2009.

9   The privatisation of air power

Peter Layton

In small and irregular wars, the state has traditionally had a monopoly on the employment of air power. This fundamental characteristic has now been overturned. With the rise of low-cost, commercial off-the-shelf drones, armed non-state actors can now employ air power in irregular wars of all types. This development was initially convincingly demonstrated with the use of such drones offensively in Iraq by the Islamic State of Iraq and Syria (ISIS) during 2014–2017. Since then other non-state groups have taken up the concept with the result that air power has now been privatised. The state's exclusive use of air power has now been broken.

So-called hobbyist drones became widely available as a mass consumer product from the mid-2000s.[1] They are easily assembled, readily flown with little training, require no infrastructure, and have rapidly declining purchase costs—today in the high hundreds of dollars. As air vehicles, the drones carry only diminutive payloads and have a limited range and endurance. However, their electronic fits can be remarkably sophisticated and include high-definition imagery sensors, real-time video transmission capability, GPS navigation systems and optimised flight control software. Civil aviation authorities typically define such drones as model aircraft as they weigh less than 25 kilograms.[2]

Armed non-state actors have long been interested in model aircraft; for example, the Aum Shinrikyo terrorist group trialled radio-controlled helicopters in the early 1990s.[3] Insurgents have also used crewed aircraft occasionally, including during the Nigerian civil war and the Sri Lankan conflict.[4] The new technology of consumer drones allowed the two streams to combine, meaning armed non-state actors could now use air vehicles in combat operations. ISIS first appreciated the tactical possibilities, with the group demonstrating the ability to use consumer drones in combined arms teams, to employ them in mass and to continuously innovate in response to drone countermeasures.

This chapter initially discusses the factors that made this revolution possible and the early employment of drones by ISIS. The second section considers the potential implications of these developments for future defence force operations in small and irregular wars. The conclusion reflects on changes that may be needed to generic air power theory.

DOI: 10.4324/9781003230656-13

ISIS' use of consumer drones

Non-state armed groups do not generally develop equipment specific to their purposes as they are resource-constrained. One option is to capture military equipment from government forces. Another is to exploit readily available commercial products, often modifying these for their needs. Contemporary examples of the latter include using smartphones for communications, garage door opening devices for improvised explosive device triggering, and social media software for recruitment and propaganda. Consumer drones fall into this category.

The modern consumer drone incorporates various new technologies arising from the information technology (IT) revolution and advances in plastic manufacturing processes. The IT revolution provided low-cost processors that allowed people to easily fly complex drone designs (e.g. quadcopters) from smartphones and tablets using local ad hoc wireless networks while operating miniaturised high-definition video cameras. New plastic moulding techniques complemented this. Previously, model aircraft were sold in complicated kit form that generally involved working in wood and metal and required long build-times. This barrier to entry was almost completely removed by introducing plastic moulded air vehicles that came in either ready-to-fly or almost ready-to-fly form.[5] The combination of low-cost consumer electronic packages and plastic moulded air vehicles allowed mass production that progressively drove down prices and made consumer drones widely affordable.

In recent years, the commercial off-the-shelf drones drone market has bifurcated into enterprise drones and consumer drones. Enterprise drones used by commercial entities are now the fastest-growing drone segment with growth high in construction, built inspection and agriculture uses. In contrast, the consumer drone market is rapidly maturing and expected to plateau by about 2025, although still with large numbers of drones produced annually. In 2019 alone, almost six million consumer drones were manufactured. Enterprise drones are more sophisticated but also more costly than consumer drones.[6] ISIS pioneered the use of consumer drones in warfare. Their low cost and easy availability globally means this type of drone is likely to remain preferred by armed non-state groups

ISIS first became interested in small consumer drones in 2013, and by 2015, it had formed a specialist drone unit.[7] Various types of rotary and fixed-wing drones were then acquired and trialled to determine the most useful kinds. The most common type of ISIS drones appeared to be of Chinese origin, particularly the rotary-wing DJI Phantom quadcopter drones and fixed-wing X-UAV Talon and Skywalker X8 FPV types.[8] The pervasiveness of Chinese drones is unsurprising, given that Chinese firms dominate the drone market. DJI is the world's largest consumer drone manufacturer by a considerable margin with its Phantom quadcopter called the 'Model T' of the industry.[9]

ISIS also acquired GoPro cameras, memory cards, GPS units, digital video recorders, and extra propeller blades to support the various drones. As it had done with its work to improve rockets and mortars, ISIS also modified the drones to suit its purposes. Conscious that drone wireless communications could be intercepted, the group investigated encrypted video transmitters and receivers. Similarly, desirous of increasing drone range, consideration was given to acquiring the long-range radio control relay systems produced by Foxtech.[10]

ISIS is believed to have begun using drones operationally in 2014. The first definite indication came in August 2014 when footage from an ISIS DJI Phantom quadcopter that undertook a reconnaissance mission to survey and map a Syrian Air Base near Raqqa was posted on YouTube. The mission detected several entry paths that were later used by suicide bombers to capture the base. In the battle for Kūbānī (also Kobanê or Kobanî), DJI quadcopters were used again to locate artillery and mortar targets and correct their fall of shot. The quadcopters also undertook dedicated information operation missions to provide high-quality footage for propaganda videos.[11]

ISIS undertook similar drone missions across 2015, but as US forces entered the conflict, ISIS drone operators began to come under attack from USAF piloted and unpiloted aircraft. At least eight strikes on ISIS drone support vehicles and associated personnel occurred, and ISIS drones were also shot down by small arms fire. Syrian rebel group The Levant Front (or *al-Jabhat aš-Šāmiyya* claimed they shot down an X-UAV Talon fitted with cameras undertaking a reconnaissance mission. Al-Qaeda and the Iraqi police also made similar claims concerning downing ISIS drones.[12]

In 2016, ISIS began using drones it had modified to employ weapons. On at least three occasions in the second half of 2016, ISIS used drones modified to be improvised explosive devices. One of these attacks succeeded, killing two Kurdish fighters and wounding two French Special Force personnel. The drone had been modified to conceal explosives in its frame, which detonated when the drone was inspected after being taken to a Kurdish facility.[13]

In the battle for Mosul (October 2016–July 2017), ISIS employed drones extensively for various types of missions as it endeavoured to prevent Mosul from being recaptured by Iraqi and Peshmerga forces. ISIS forces proved ineffective in the city's surrounding rural areas and on its outer edges but mounted a protracted and taxing defence of Mosul's inner-city areas. One of ISIS' signature weapons was the suicide vehicle-borne improvised explosive device (SVBIED): a car loaded with explosives detonated when the driver ran the vehicle into a designated target such as tanks, Humvees, and static checkpoints. Such attacks were coordinated with other ISIS operations in a combined force action to overrun or turn Iraqi Security Force (ISF) positions.

The ISF blockaded streets with rubble or wrecked vehicles to counter the SVBIED threat in Mosul's urban areas. In turn, ISIS developed effective tactics to overcome these blockades, including integrating SVBIED with drone missions. The drones—usually DJI quadcopters—provided real-time reconnaissance video, which ISIS used to guide the SVBIEDs through narrow side streets to avoid checkpoints and defensive roadblocks, thereby ensuring they survived long enough to attack their selected targets. The quadcopter high-definition video imagery of the attack was then quickly uploaded to the internet for propaganda purposes. ISF and its supporting coalition forces responded to these tactics by launching air strikes (including from orbiting USAF Predator drones) against the on-ground ISIS drone controllers. In turn, ISIS countered this response by mobilising drone controllers, who would move around the city using motorcycles.[14]

However, the major drone innovation by ISIS in the Mosul battle was its use of weaponised drones.[15] ISIS modified DJI Phantom quadcopter drones to carry and drop small munitions such as grenades and 40 mm mortar shells. These munitions were themselves often modified by adding fins to stabilise their fall. The drones were designed to use high-definition cameras, hover overhead a stationary target, and provide a fully stabilised platform for accurate freefall weapons delivery.[16] In Mosul, ISIS also opened small factories to undertake quantity production of the modified drones and munitions.

Attacks by these weaponised drones began in early November 2016. They quickly escalated in scale, and by February 2017, some 70 drone attacks were reported in a single 24-hour period, with 12 counted overhead simultaneously.[17] The drones hunted tactical targets such as groups of ISF personnel, stationary Humvees, and parked tanks together with humanitarian aid distribution centres in an attempt to disrupt ISF stabilisation efforts. Although each attack caused only limited damage, the persistent harassment by the so-called killer bees, which operated during the day and night, adversely impacted morale. At one point, the ISF offensive to retake Mosul almost stalled.[18]

US Special Operations Command (SOCOM) units were advising and support-ing ISF at the time. The SOCOM head, General Raymond Thomas, noted of this period that the

> most daunting problem was [that ISIS] . . . for a time, enjoyed tactical supe-riority in the airspace under our conventional air superiority in the form of commercially available drones and . . . our only available response was small arms fire.[19]

Overtime, ISF and US forces steadily reduced the threat by deploying counter-drone jamming systems, jamming command links and video feeds, and striking ISIS drone factories, launch sites and associated vehicles. At the end of the battle, ISF Lieutenant General Abdul-Amir Yarallah claimed Iraqi units had shot down some 130 ISIS drones.[20]

In their three years of operational use, ISIS used drones for three principal roles: intelligence, surveillance, and reconnaissance (ISR); information operations; and offensive operations. Being consumer drones, they were readily defeated when friendly forces were aware of their presence, making the principal threat the use of drones en masse and the innovative ways the drones were modified and employed. Captured ISIS documents suggested an interest in acquiring fixed-wing drones for longer-range missions and in air-delivering chemical weapons.[21]

Implications for future Australian Defence Force operations

ISIS' success suggests future insurgencies may also feature consumer drones undertaking similar roles to those ISIS employed and potentially new ones. In that regard, the Mosul battle takes on added importance.

With large-scale population shifts globally from rural areas into cities, many armies now expect future wars to be urban.[22] In this regard, a post on an Australian Army website declared that in urban warfare, armed airborne ISR is now 'the king of the urban battlefield . . . An Army without organic airborne armed ISR will be at a severe disadvantage on a contemporary urban battlefield'.[23] Although this statement referred to the USAF's Reaper remotely piloted air vehicles, the principal attributes the author determined to be needed are also those of ISIS' modified consumer drones: persistent stare from above; target identification capability; the ability to coordinate with artillery to guide indirect fire accurately; and, if necessary, the ability to attack with a low-collateral weapon.[24] In Mosul, ISIS had its own organic 'king of the urban battlefield'. Future non-state actors will also.

This privatisation of air power continues with the Ukraine War offering some further signposts. While ISIS was dropping bombs in Iraq, Ukrainian non-state actor Aerorozvidka was also developing armed drones this time to attack Russian separatist forces in Eastern Donbas; this work involved volunteers from numerous IT companies and was funded by crowdsourcing.[25] By the time of the Russian invasion in February 2022, Aerorozvidka had moved beyond DJI drones to using drones of its own design patched together from readily available commercial components. Its armed R-18 quadcopter had a range of four kilometres, an endurance of 40 minutes and dropped five-kilogram bombs. This drone was used in a combined arms fashion with quad-bike mounted teams to attack Russian supply convoys using rural roads.[26]

A major implication of this increasing number of armed non-state actors using drones is that both sides will employ air combat capabilities in future small and irregular wars. Only government forces employed air power in previous conflicts of this nature, but now insurgents may also have these capabilities. Western land forces have not faced a hostile air environment since the Korean War of the early 1950s, but this situation no longer exists. The low cost of drones and their ready availability means that at any time, an insurgency could use drones to gain a tactical advantage, whether in terms of ISR, information operations or attack. Western forces involved in counter-insurgency operations will need to deploy appropriate air defence systems to counter the likely ongoing drone threat.[27]

Moreover, as consumer drones become more technically advanced, they are expected to become more autonomous—making jamming less effective—and feature multivehicle control, allowing swarming attacks. Massed attacks using autonomous drones would be readily detectable but be hard to defeat, with the danger of the facility or airbase under attack being overwhelmed. Noting such issues, Kelley Sayler warns that such drone swarms 'could enable a . . . non-state actor . . . to achieve capability overmatch against the United States.'[28]

The enduring nature of the insurgent drone threat and possible future swarm tactics suggests that active air defences alone are potentially insufficient. Some form of passive defences for fixed bases may be required to counter drone attacks using small munitions and impede drone ISR missions surveying facilities for a future insurgent attack. Camouflage measures may need to be taken and, for tactically deployed forces, some simple passive defence measures might also be useful. For example, many consumer drone video cameras cannot penetrate smokescreens; these might be quickly laid to cover the short periods insurgent drones can loiter overhead.

Implications for air power theory

The rise of hostile consumer drones also has implications for air power theory. Firstly, the most significant implication is that gaining and maintaining air superiority is now a crucial mission across all forms of conflict. No longer can air superiority be assumed in lesser wars. More pointedly, the techniques used to gain air superiority in a conventional war may be less productive in these small or irregular wars. As General Thomas observed about the Mosul battle: 'At one point there were twelve "killer bees" . . . right overhead and underneath our air superiority'.[29] ISF and coalition ground forces in Mosul did not lack friendly air support with a large, diverse force of fast jets fighters available. However, the force was ill-suited for the task, as pop-up drones presented unusual air targets which were hard to engage consistently and quickly.

Secondly, consumer technology can now provide useful air power. ISIS drones acted as force multipliers to support and direct ISIS ground force attacks in a manner well-known to air power theorists. In previous counter-insurgency wars, simpler lower-technology aircraft have at times proven more useful than more advanced ones. For example, in the Vietnam War, some USAF F-100 Super Sabre squadrons traded in their supersonic jets for older, piston-engine A-1 Skyraiders that were more efficacious in providing on-call close air support. Today, commercial drones have taken air power another step down the technology rung and into consumer products.

Thirdly, the future air power capabilities of non-state actor drones will be driven by consumer demand. Air power has usually been advanced and shaped by the demands of armed forces, although there have been some exceptions. For example, the demands of airline passengers and companies in the interwar period led to air transport aircraft that could fly faster and higher than contemporary fighters. This example also highlights the differences with the interwar period: the numbers of airliners were in the hundreds, whereas today's drone market is annually in the millions.

The vast global market for drones and the technology used will evolve in a direction commensurate with the amount its customers are willing to spend. While some very large firms such as DJI are involved in mass drone production, many smaller companies are also developing new applications, flight control software, and payloads. Future drone innovations will be developed from below and, when proven, drawn up into mass production. Regardless of who develops it, future consumer drone technology will continue to leverage the commercial IT industry with its large investments. Consumerism in a very real way will drive non-state actor air power.

Lastly, air power has always been somewhat elitist. It is normally wielded by large, well-financed government organisations composed of numerous highly skilled people. Drones overturn this model, as they are affordable and can be flown and operated by most. Drones disperse air power well beyond the confines of the traditional entities that have employed it across the first century of powered flight. Drones have now moved air power away from government ownership and control and passed this to the people. Consumer drones have privatised air power.

Notes

1 A four-part drone classification system is developed in Kelly Sayler, *A World of Pro-liferated Drones: A Technology Primer* (Washington, DC: Center for a New American Security, 2015).
2 An example is the US FAA which mandates less than 55 lbs (25 kg). See: Federal Aviation Administration, *Advisory Circular No. 91–57a: Model Aircraft Operating Standards* (Washington, DC: US Department of Transportation, 2 September 2015), 2.
3 Larry Friese, N. R. Jenzen-Jones, and Michael Smallwood, *Emerging Unmanned Threats: The Use of Commercially-Available UAVs by Armed Non-State Actors* (Perth: Armament Research Services, 2016), 36, http://armamentresearch.com/wp-content/uploads/2016/02/ARES-Special-Report-No.-2-Emerging-Unmanned-Threats.pdf.
4 The Biafran insurgents in the Nigerian civil war employed a variety of different aircraft for various purposes; see: Victor Flintham, *Air Wars and Aircraft: A Detailed Record of Air Combat, 1945 to the Present* (New York: Facts on File, 1990), 99–103. The Libera-tion Tigers of Tamil Eelam (LTTE) acquired several aircraft during the long-running Sri Lanka conflict and used them quite aggressively; see: Sergei Desilva-Ranasinghe, "Insurgent Airpower and Counter Terrorism," *Defense Asia Review* 3, no. 5 (2009): 25–28.
5 Friese, et al., *Emerging Unmanned Threats,* 30–32.
6 Dario Constantine, "The Future of the Drone Economy," *Levitate Capital*, Decem-ber 2020, 107, https://levitatecap.com/levitate/wp-content/uploads/2020/12/Levitate-Capital-White-Paper.pdf.
7 Don Rassler, Muhammad Al`Ubaydi, and Vera Mironova, "The Islamic State's Drone Documents: Management, Acquisitions, and DIY Tradecraft," *Combating Terrorism Center,* 31 January 2017, https://ctc.usma.edu/ctc-perspectives-the-islamic-states-drone-documents-management-acquisitions-and-diy-tradecraft/.
8 Friese, et al., *Emerging Unmanned Threats*, 43.
9 Jack Nicas and Colum Murphy, "Who Builds the World's Most Popular Drones?" *The Wall Street Journal*, 10 November 2014, www.wsj.com/articles/who-builds-the-worlds-most-popular-drones-1415645659.
10 Rassler, et al., "The Islamic State's Drone Documents".
11 Don Rassler, *Remotely Piloted Innovation: Terrorism, Drones and Supportive Technol-ogy* (West Point: Combating Terrorism Center, October 2016), 35–36.
12 David Hambling, "ISIS Is Reportedly Packing Drones with Explosives Now," *Popu-lar Mechanics*, 16 December 2015, www.popularmechanics.com/military/weapons/a18577/isis-packing-drones-with-explosives/.
13 Rassler, *Remotely Piloted Innovation*, 37–38.
14 Michael Knights and Alexander Mello, "Defeat by Annihilation: Mobility and Attri-tion in the Islamic State's Defense of Mosul," *CTC Sentinel* 10, no. 4 (April 2017): 3.

15 Seth Robson, "Islamic State Attack Drones Pose Threat to Iraqi Troops, General Says," *Stars and Stripes*, 1 February 2017, www.stripes.com/news/islamic-state-attack-drones-pose-threat-to-iraqi-troops-general-says-1.451913#.Wbm4ENMjGV6.

16 Ben Watson, "The Drones of ISIS," *Defense One*, 12 January 2017, www.defenseone.com/technology/2017/01/drones-isis/134542/.

17 Knights and Mello, "Defeat by Annihilation," 5–7.

18 Howard Altman, "Tale of Two Drones: ISIS Wreaked Havoc Cheaply, Tampa Meeting Showcases State of the Art," *Tampa Bay Times*, 16 May 2017, www.tampabay.com/news/military/tale-of-two-drones-isis-wreaked-havoc-cheaply-tampa-meeting-show cases/2324138.

19 David B. Larter, "SOCOM Commander: Armed ISIS Drones Were 2016's 'Most Daunting Problem," *DefenseNews*, 16 May 2017, www.defensenews.com/digital-show-dailies/sofic/2017/05/16/socom-commander-armed-isis-drones-were-2016s-most-daunting-problem/.

20 Karzan Sulaivany, "Offensive in Mosul Killed over 25,000 IS Militants: Commander," *Kurdistan 24*, 16 July 2017, www.kurdistan24.net/en/news/6efb35b8-4b32-4536-8a84-fd89aaa7a589.

21 Rassler, et al.; "The Islamic State's Drone Documents".

22 For example, see Modernisation and Strategic Planning Division – Australian Army Headquarters, *Future Land Warfare Report* (Canberra: Commonwealth of Australia, April 2014).

23 LTCOL N, "Immediate Lessons from the Battle of Mosul," *Australian Army Research Centre*, 25 June 2017, https://researchcentre.army.gov.au/library/land-power-forum/immediate-lessons-battle-mosul.

24 LTCOL N, "Immediate Lessons from the Battle of Mosul," *Small Wars Journal*, 26 June 2017, https://smallwarsjournal.com/blog/immediate-lessons-from-the-battle-of-mosul.

25 Christian Borys, "Crowdfunding a War: Ukraine's DIY Drone-Makers," *The Guardian*, 24 April 2015, www.theguardian.com/technology/2015/apr/24/crowdfunding-war-ukraines-diy-drone-makers.

26 Ahmad Ghayad, "This is How Ukraine is Building DIY Drones to Destroy Russian Vehicles!" *Engineerine*, 13 April 2022, www.engineerine.com/2022/04/ukraine-is-building-diy-drones-to-fight.html.

27 There are many possible systems. See, for examples Guy Martin, "Attacking the Killer Drones: Counter-UAV Systems," *Defence Review Asia* 11, no. 4 (August 2017), 18–22.

28 Sayler, *A World of Proliferated Drones,* 29.

29 Larter, "SOCOM Commander".

References

Government sources

Federal Aviation Administration. *Advisory Circular No. 91–57a: Model Aircraft Operating Standards*. Washington, DC: US Department of Transportation, 2 September 2015.

Modernisation and Strategic Planning Division – Australian Army Headquarters. *Future Land Warfare Report – 2014*. Canberra: Commonwealth of Australia, April 2014. https://researchcentre.army.gov.au/library/other/future-land-warfare-report-2014.

Published sources

Desilva-Ranasinghe, Sergei. "Insurgent Airpower and Counter Terrorism." *Defence Review Asia* 3, no. 5 (2009).

Flintham, Victor. *Air Wars and Aircraft: A Detailed Record of Air Combat, 1945 to the Present*. New York: Facts on File, 1990.

Friese, Larry, N. R., Jenzen-Jones, and Michael Smallwood. *Emerging Unmanned Threats: The Use of Commercially–Available UAVs by Armed Non-State Actors*. Perth: Armament Research Services, 2016. http://armamentresearch.com/wp-content/uploads/2016/02/ARES-Special-Report-No.-2-Emerging-Unmanned-Threats.pdf.

Knights, Michael, and Alexander Mello. "Defeat by Annihilation: Mobility and Attrition in the Islamic State's Defense of Mosul." *CTC Sentinel* 10, no. 4 (April 2017): 1–7.

Martin, Guy. "Attacking the Killer Drones: Counter UAV Systems." *Defence Review Asia* 11, no. 4 (August 2017): 18–22.

Rassler, Don. *Remotely Piloted Innovation: Terrorism, Drones and Supportive Technology*. West Point: Combating Terrorism Center, October 2016.

Sayler, Kelly. *A World of Proliferated Drones: A Technology Primer*. Washington, DC: Center for a New American Security, 2015.

10 Hypersonic propulsion as an air power disruption or disturbance

Michael Spencer

The Royal Australian Air Force's (RAAF) Air and Space Power Centre (ASPC) (formerly known as the Air Power Development Centre) has searched forecasts for change drivers and disruptions that are currently recognised by analysts employed in the military, business, science, and technology, which are relevant to prompt disruptive thinking when considering the necessary future improvements to contemporary air power. In the RAAF's *Beyond the Planned Air Force*,[1] RAAF aviators are challenged to both identify and explore technological, societal, and environmental factors that can both disrupt and drive changes that may shape the ability of the RAAF to achieve its mission: to provide air power in support of Australia's national interests across all domains. *Beyond the Planned Air Force* establishes a period for considering future air power beyond the timelines within current plans for force designs and the Integrated Investment Program that guides future capability investments and organisational structuring.[2]

In researching future disruption to the application of air power, it is essential to adopt a standard definition for 'disruption' that allows us to recognise when something has the potential to change the viability of a technology option and its benefits when used in air power. Whilst there is no previously agreed definition available for the context in this discussion, a review of the extant literature focuses on the importance of disruption to continuing business viability and change management.[3] This review suggests an appropriate meaning for 'disruption': a person or thing that prevents the system, process, or event from continuing as usual or as planned. If seen in advance, a disruptive technology might be considered a threat to the continuing viability of the business or the continued effectiveness of employing a particular technology, which would, in turn, necessitate a business change to mitigate that threat.

In recent years, hypersonic air power has become a topic that the ASPC has reviewed in detail. Collaboration with 'air power practitioners',[4] commentators, and academics working in the field has afforded the Centre a greater understanding of hypersonic air power.[5] Furthermore, hypersonic technology has become a popular research theme globally, both academic and military. The hypersonic manoeuvre may enable air attack systems to better penetrate and survive against contemporary air defence system designs. In reviewing the status of hypersonic

DOI: 10.4324/9781003230656-14

air power as a disruptor, this chapter aims to provide a helpful knowledge baseline when considering whether introducing hypersonic air power into the future battlespace could disrupt contemporary air power concepts and how they are applied in the future.

The Australian trajectory into hypersonic propulsion

Australian military planners understand the benefits of having a speed advantage to reduce the time available for an adversary to understand and respond to actions in the battlespace. The quest for faster machines has inspired research into hypersonic propulsion and controlled flight at hypersonic speeds. However, the gap between the theory and reality has resulted in very long, drawn-out, and expensive research programmes by different countries, including Australia.[6]

The mid-20th century Cold War between the Soviet Union and the United States drove an arms race that inspired the rapid developments of new high-speed technologies. The tensions escalated efforts to develop missiles to fly suborbital trajectories at hypersonic speeds over intercontinental distances and precisely deliver a warhead. The successful developments of ballistic missiles resulted in spinoff research and technology applications exploring the potential benefits of high-altitude and high-speed trajectories.

During this period, Cosmonaut Yuri Gagarin was the first to fly at hypersonic speed as his freefalling orbital spacecraft re-entered the Earth's atmosphere. NASA developed the X-15 programme as the first aircraft to perform manoeuvres at hypersonic speeds. Hypersonic flight had finally realised a step-change in capabilities. However, it had been a long-time coming and still has a long way to go before practical military applications can be realised.

Australian Professor Raymond Stalker pioneered the supersonic combustion ramjet (aka scramjet), a necessary enabler towards achieving air-breathing hypersonic propulsion. From 1988 to 1993, as Australia's first Professor of Space Engineering, he led the NASA-supported Centre for Hypersonics at The University of Queensland, which conducts world-leading hypersonic research.[7]

The Australian Department of Defence has collaborated with universities, commercial, and international partners. Collaborative hypersonic research programmes include the 2002 HyShot flight test of a scramjet engine; the 2012 Hypersonic International Flight Research Experimentation (HIFIRE) programme;[8] the 2013 Scramspace hypersonic scramjet engine;[9] and the 2020 Southern Cross Integrated Flight Research Experiment (SCIFiRE) cooperative programme to flight test prototype hypersonic missiles.[10]

The Australian Government supports hypersonic research programmes to better position the Australian Defence Force to respond to future threats.[11] In its 2020 Defence Strategic Update, the government reinforced this position with plans to include the 'development of advanced air-to-air and strike capabilities with improved range, speed, and survivability, potentially including hypersonic weapons'.[12]

Hypersonic disruption

Propulsion systems are currently being developed for air vehicles to achieve speeds beyond Mach 5—the threshold for hypersonic speed. Hypersonic manoeuvre is being pursued by major military powers such as China, Russia, and the United States to develop new propulsion technology and realise a new operational advantage in the future battlespace. Current concerns related to hypersonic speeds and manoeuvres focus primarily on how hypersonic technology affects the responsiveness of contemporary command and control arrangements and air defence systems designed to counter attacks from traditional air vehicles designed to travel at subsonic or supersonic speeds.[13] Hypersonic vehicles are expected to be capable of traversing long distances faster than current air defence systems can react and activate countermeasures.[14]

The potentially disruptive attributes of hypersonic air power are its speed and manoeuvre and the ability to travel at hypersonic speed and change trajectory while doing so. Hypersonic vehicles introduce a new threat that can manoeuvre and reach its objective faster than a traditional air defence system can detect and respond to an air threat. Although a conventional air defence system might gain an early detection against an incoming hypersonic air threat, it is expected that the time taken to process the threat and decide a response is likely to be too slow to deliver an effective counteraction. In this way, hypersonic propulsion can defeat the decision cycles used to determine operational responses in traditional air defence systems, which are typically based on human-speed decision processes and drive a need to develop machine-speed decision-making systems.

Hypersonic air attack

Long-range attack weapons that can traverse intercontinental distances and may be launched from the safety of a homeland base or an otherwise benign or unknown launch point that is not within reach of air defence systems represent a significant foundation of the national security strategies of the world's nuclear powers. For example, during the Cold War, the United States and the USSR developed and integrated the inter-continental ballistic missile (ICBM) into strategic air power to complement piloted long-range strategic bombers. As a result, the ICBMs became the long-range attack weapon of choice to ensure nuclear deterrence between these two nuclear super powers; travelling at speeds more than Mach 20 on re-entry to the atmosphere, these missiles were 'essentially immune' to defensive action.[15]

In designing a precision guidance and control system for a hypersonic air vehicle, technological limitations meant that Cold War-era ICBMs relied on a freefall ballistic trajectory to reach their planned objective. There were inherent inaccuracies in the precision of the ballistic freefall trajectory of ICBMs. Still, these were made irrelevant by using large and high-yield conventional or nuclear warheads. The damage radius from a large warhead can offset missile inaccuracy and, conversely, a very accurate weapon can reduce the damage radius.[16] The operationalisation of hypersonic propulsion in controlled flight would enable the delivery of smaller conventional warheads in long-range precision attacks. The smaller warhead damage effects would be compounded by the kinetic energy from missile debris travelling at hypersonic speeds. When considering the variables in missile designs, a smaller warhead could potentially reduce the overall missile size to improve its aerodynamic efficiency, low observability, and stealth and increase its performance range with the same fuel load.

The applicable laws of physics for the performance of an ICBM can be simple. To fly a payload from the launch point over the required distance to the impact location requires the measured application of rocket power (i.e., based on mathematical predictions of rocket force and engine duration), the steering of the missile onto a trajectory calculated with the altitude, direction, and range required to reach the ground objective. First, the rocket accelerates the payload to hypersonic speed in a ballistic trajectory. Once the propellant is spent, the rocket boosters cease to function and are ejected as debris, which falls in an uncontrolled descent. Next, the hypersonic payload continues along its ballistic trajectory, which can take it beyond the atmosphere until it traverses the maximum height of its ballistic trajectory and begins an uncontrolled freefall under the effects of gravity, re-entering the atmosphere at a hypersonic speed and falling ballistically onto its target location.[17]

If the type of rocket vehicle, launch angle, and ballistic trajectory can all be observed and measured by sensors, it is possible to estimate the trajectory. The estimated trajectory projection can then be used to predict a probable impact position for a freefalling ballistic payload, even if it is travelling at hypersonic speed. Since this design for strategic weapons has been popular with the nuclear powers, anti-ballistic detection and defence systems have since been developed and dedicated to the hypersonic freefalling ballistic payloads.

Within the scope of this chapter, anti-ballistic missile defence is not regarded as a conventional role; however, the fundamental roles currently identified for air power will continue to have merit after hypersonic air power is militarily operationalised. The typical designs for an ICBM are based on using rocket propulsion to propel a warhead onto a ballistic trajectory that may extend beyond the atmosphere and freefall through the atmosphere at a hypersonic speed. The cost implications, technical complexity, and the politico-strategic risk implications limited their employment to superpower nations and authoritarian states willing to prioritise state funds in their investment. Only a smaller number of countries could further afford the research and technology investments needed to realise the strategic defence capabilities required to detect and defend against attacking ICBMs. Typically, detecting and engaging an ICBM in the early boost phase of its ascent is optimal, while the functioning booster provides a good signature for making a long-range detection that also maximises the time available to analyse the adversary's intent and process the decisions needed and choose the best available operational response option. Once the rocket booster expires and is jettisoned, the smaller sized warhead continues to freefall ballistically at hypersonic speed.

Next-generation missile designs are currently being considered to configure a traditional rocket engine to boost a glide vehicle to hypersonic speed. Additionally, some designs for ICBM engines may be repurposed for launch from combat aircraft, warships, and truck erector launchers. The boosted glide vehicle is initially launched into an exo-atmospheric ballistic trajectory. After re-entering the Earth's atmosphere, when it reaches an altitude where the air density is adequate for aerodynamic forces to enable controlled flight, the vehicle transitions from a hypersonic ballistic trajectory to hypersonic controlled flight to continue to its objective.

However, hypersonic glide vehicles represent a potential tactical disruption to the traditional air defence systems designs due to their ability to manoeuvre in the air compared to the ballistic freefalling trajectories of ICBMs. The continuation of the flight trajectory after the ballistic freefall along a low altitude profile poses challenges to air defence surveillance and tracking sensors that have traditionally been designed to defend against hypersonic ballistic vehicles freefalling from high altitudes. Hypersonic glide vehicles that survive the ballistic freefall and commence flying at a low altitude trajectory are likely to penetrate traditional air defence systems by reducing the time available to effectively respond to and intercept the incoming hypersonic vehicle before it reaches its objective.

Hypersonic propulsion options

Hypersonic speed changes the chemistry of the airflow and the behaviour of air. Air combat systems optimised for subsonic and supersonic airflow cannot operate as intended in hypersonic airflow. Non-hypersonic designs cannot readily be extrapolated to function in a hypersonic airflow due to higher skin temperatures caused by hypersonic air friction, the hypersonic airflow disrupting the correct functioning of air-breathing engines designed for conventional speeds, and flight controls that need to be designed to be effective in hypersonic airflow. Accordingly, the hypersonic operating environment requires that hypersonic airflow be recognised as another distinct air operating environment with different characteristics from the subsonic and supersonic air environments. This necessity presents unique challenges to system designers for building new machines that can manufacture new vehicles and subsystems for propulsion, navigation, guidance, and flight controls. These subsystems will be challenged to function in the harsh conditions posed by hypersonic speeds, potentially whilst travelling in the air and outer space, and be capable of autonomously manoeuvring to make trajectories corrections over long distances to reach and cross a distant battlespace.

The propulsion system is the critical enabler for hypersonic air power. Although prototype systems for electromagnetic railguns have been observed under evaluation on American and Chinese warships, with BAE Systems receiving a USD34.5 million contract from the United States' Office of Naval Research in 2013 to develop the technology, these systems are designed for surface warfare.[18] Electromagnetic railguns operate by accelerating projectiles to hypersonic speed over tactical engagement ranges, relying on the kinetic energy of the hypersonic impact (e.g. Mach 3 to 6) to cause penetration and fragmentation damage at the target. They are not, however, considered further in this chapter.

Two basic propulsion engine designs are currently employed for propelling air vehicles over intercontinental distances and into an air trajectory travelling at hypersonic speeds similar to those of rocket-boosted freefalling ICBM warheads. The first type is the traditional rocket-boosted freefalling ballistic payload that travels at speeds starting at Mach 6 and ranging up to more than Mach 25. Rocket' boost and glide' systems employ rocket engines to boost a payload into sub-orbital or orbital trajectories for a controlled re-entry into the atmosphere and freefall along a ballistic trajectory over intercontinental distances to reach its terrestrial target. There has been a significant history in the development of technologies for ICBM and the realisation of hypersonic payloads delivered by the ICBM. The rocket engines typically only function during the ascent to boost the payload beyond the atmosphere, at which point gravity pulls the freefalling payload back to Earth. During the atmospheric re-entry, the vehicle can fall at airspeeds starting from approximately Mach 25 down to Mach 6, which causes significant aerodynamic heating and harsh structural vibrations that may cause damage or malfunctions in the vehicle and its payload.

Whereas an ICBM payload was traditionally designed to freefall ballistically along a predictable trajectory to its target objective, an aerodynamically manoeuvring vehicle that can travel at hypersonic speed offers the potential to defeat air defence systems that have been designed to counter the conventional ICBM warheads that follow a predictable freefall trajectory. For example, the Chinese DF-17, a medium-range ballistic missile, is the first designed for the People's Liberation Army Rocket Force (PLARF) that enables its deployment of a manoeuvring hypersonic glide vehicle (HGV) referred to as the DF-ZF (previously known as the WU-14). The DF-ZF HGV, it is claimed, can reach speeds between Mach 5 and Mach 10, and it is suggested that the vehicle may have been designed to carry either a nuclear weapon or precision-strike conventional warhead. Some experts anticipated the DF-ZF's entry into service as early as 2020. Still, regardless of when it becomes operational, strategic security implications of new hypersonic glide vehicle designs have been identified.[19] American advocacy group Missile Defense Advocacy Alliance notes that the DF-ZF affords the PLARF a hypersonic capability that enables missiles carrying the DF-ZF to achieve shorter flight times and perform evasive manoeuvres while travelling at hypersonic speeds. This capability would pose a significant threat to nations in the region, including South Korea and Japan, which have invested heavily in ballistic missile defence systems in recent years. Current missile defence systems would be limited in their ability to intercept such missiles, with the DF-ZF design capable of carrying nuclear and conventional payloads.[20]

The second available propulsion system is the Supersonic Combustion Ramjet (scramjet), which can travel at airspeeds in excess of Mach 6. The scramjet is a ramjet variant, an air-breathing jet engine that lacks a rotary compressor. Unlike the ramjet, in which air is decelerated to subsonic velocities before combustion occurs in the engine, the air remains at supersonic velocities in the standard ramjet as the rammed air is still travelling at supersonic speed when it is mixed and combusted with the propellant to generate thrust.[21] The carefully and precisely formed shape of the vehicle forebody, located ahead of the engine air intake, compresses the airflow before it enters the combustion chamber. Fuel is then injected, and combustion occurs in supersonic airflow. Compared to a turbine engine, the ramjet and scramjet have a technological advantage: they do not require moving parts to accelerate the airflow and generate thrust to propel the engine in the opposite direction. However, they also need to be integrated with complementary non-hypersonic propulsion systems to accelerate them from zero to hypersonic speed.

The scramjet is usually integrated into the design of the air vehicle body, symmetrically around the central axis of the vehicle design because hypersonic airflow around non-symmetrical vehicle designs is prone to causing instability in flight.[22] The US Air Force X-51A Waverider is an example of such an aircraft. Designed by Boeing, the Waverider, so named because it is designed to ride its shockwave at hypersonic speed, is an uncrewed scramjet demonstration aircraft intended as a missile-size demonstration aircraft. Using JP-7 aviation fuel, the X-51A is launched from a B-52 Stratofortress bomber to demonstrate scramjet operation from Mach 4.5 to Mach 6. However, although four were built for the US Air Force, they were not designed to be a prototype for a weapons system, only to 'pave the way' for future hypersonic technology development.[23]

A third propulsion system design represents a hybrid of these two systems, combining air-breathing and non-air breathing systems at airspeeds from Mach 3 to in excess of Mach 10. This design represents a system that combines an air-breathing system for operating at low speeds in oxygen-rich low altitudes and a ramjet or rocket for operating at high speeds in the high altitudes of the rarefied and zero atmosphere. A combination of discrete systems, such as a launcher aircraft, a rocket or ramjet booster, and an onboard scramjet engine, is needed to accelerate a payload from zero to hypersonic speed. A hybrid engine system, such as a combination design with a rocket and scramjet, might realise a single discrete system that could accelerate a payload from zero to hypersonic speeds. For instance, the British concept designs for a single-stage-to-orbit spaceplane are configured with an alternating combined-cycle known as the Synergetic Air-Breathing Rocket Engine (SABRE) propulsion system. The hydrogen-fuelled hybrid propulsion engine uses the atmospheric oxygen for combustion during take-off and acceleration to Mach 5.4 at high altitudes within the atmosphere. When it reaches higher altitudes, the engine switches to an internal liquid oxygen supply to combust with the rocket engine to enable propulsion beyond the atmosphere.[24]

Impacts of hypersonic manoeuvre on contemporary air power

As its basis for the successful application of air power, the Australian air power doctrine identifies three fundamental characteristics to guide the RAAF's functions for organising capabilities and defining preparedness levels to provide effective air power for the joint force. The key air power characteristics described in the RAAF doctrine are reach, perspective, and responsiveness.[25] They can be used to steer the development of air combat systems with measurable advantages over an adversary.

Speed and range are key attributes of combat power employed in or over the 'air domain'.[26] Air weapons that have been designed to exploit aerodynamic forces generated with air have been innovated and iteratively designed with capability improvements by flying increasingly faster and further to counter or survive iterative improvements in enemy air defences and countermeasures. Thus, the continual pursuit of air combat advantages has led research and development to seek to realise hypersonic propulsion systems that are affordable and viable for air combat applications in future air power. In this light, the realisation of hypersonic speed can be viewed as a tactical disruption to the traditional thinking about the speed, reach, and altitudes for air combat systems that air power practitioners can expect to be employed or encountered in the future air domain.

The ability for air vehicles to now transit and manoeuvre within the air domain at hypersonic speeds will disrupt tactical concepts that have traditionally relied on designing conventional tactical air defence systems that are baselined against subsonic and supersonic air vehicles presenting the worst-case threat scenarios. Contemporary air defence systems are not expected to have been designed to detect and respond as effectively against new hypersonic air vehicles compared to defending against traditional threats travelling at subsonic and supersonic speeds. A hypersonic missile can travel 1,000 km in about ten minutes.[27] One of the many performance indicators for an air defence system is the minimum time needed to satisfactorily detect and assess a flight track, decide, and execute a response option, and guide the missile intercept to the incoming threat. The speed of the hypersonic air threat significantly compresses the time that it is available to be detected and intercepted, degrading the effectiveness of traditional air defence systems.

Hypersonic air power in a future air domain

The introduction of hypersonic manoeuvring air vehicles in the future battlespace is likely to supersede the traditional approach to planning an air campaign. Traditionally, air campaigns have been planned such that they begin with an operational phase in which air combat elements secure control of the air before embarking on follow-on phases with other air combat and support elements to deliver their desired effects into the battlespace safely. The utility of an air attack vehicle that can manoeuvre at hypersonic speeds is grounded in its likely ability to penetrate contemporary air defence systems, which have traditionally been designed to counter air threats travelling at subsonic and supersonic speeds as the basis for their design and performance. This ability enables such attack vehicles to reach their battlespace objective without necessarily needing to follow a longstanding traditional model to invest efforts into controlling the air as a prerequisite for assuring the safety and security of other types of air missions within that same airspace. Suppose we consider the effect that hypersonic speed can have on conventional approaches to employing air power. It becomes clear:

a.　Control of the air missions, as the traditional prelude to executing the main effort in a full-scale air operation requiring separate tasks by different systems to contribute specialised efforts, may no longer be considered a necessary prerequisite for hypersonic air missions.

b.　Strike missions may be more effective because the hypersonic air vehicle is likely to penetrate and survive through an adversary's defended area by overwhelming the adversary's decision cycle. The hypersonic warhead may reach its mission objective before a traditionally designed air defence system can effectively detect and respond to a hypersonic attack.

c.　With future designs for hypersonic long-range airlift vehicles, air mobility will likely enhance mobile forces' global reach and responsiveness deployed from their parent bases located on distant home soil.

d.　Air intelligence and ISR missions will continue to be able to exploit air for its unique perspective of the battlespace and beyond.[28] Air intelligence may exploit the improved survivability enabled by hypersonic air propulsion and manoeuvring to reach far and rapidly deploy one or more sensors into and over the battlespace. Additionally, the efforts that miniaturised the technology to realise space missions with microsatellites, including long-range sensors and communications subsystems, will also benefit the designs for hypersonic air intelligence and ISR missions.

Similarly, hypersonic manoeuvres will compress the traditional models used for decision cycles that have been used as the basis for designing air surveillance and defence systems to counter subsonic to supersonic tactical attack systems. Consequently, from the opponent's perspective of providing an air defence system against possible hypersonic air attacks, air defences might be affected as follows:

a.　An opponent's investments into defensive counter-air capabilities to secure the airspace and maintain control of the air will be disrupted by hypersonic air power. Contemporary air defence systems will need improvements to command and control systems, including automation to enable machine-speed decisions to recognise and execute a response to hypersonic air vehicles at speeds much faster than is possible with human decision-makers.

b.　Strike missions by adversary hypersonic air vehicles will penetrate airspace defended by current standard systems without contest. Traditionally equipped forces will need to consider using offensive actions as a defensive countermeasure. For instance, suppose the traditional campaign approach to firstly secure control of the air over the battlespace is not practical with contemporary forces. In that case, this may make it necessary to consider 'left-of-launch strikes' (i.e., taking offensive action to defeat an ICBM threat before it is launched) to disrupt an adversary's hypersonic mission before it begins pre-emptively. In addition, task force commanders may need to consider increasing the resilience and robustness of bases to support a deployed expeditionary force with better air mobility for rapid and frequent relocations, increasing its survivability against hypersonic attacks planned against fixed locations.

c.   Air mobility with a future designed hypersonic long-range airlift vehicle that delivers payloads of miniaturised uncrewed air combat vehicles that deploy and operate in the battlespace as a swarm might overwhelm friendly early warning and ISR sensors and defences optimised for defending against incoming attacks by fewer and larger-sized systems that are easier to detect.

d.   Air intelligence and ISR systems will need to be designed and oriented to investigate earlier in the threat timeline to look further 'left-of-launch' for warnings and indicators to increase the time available for the decision process to detect, comprehend and decide a response. In addition, long-range surveillance systems may need to be improved with space-based systems and international partnerships to support the forward deployment of ground-based ISR sensors closer to an adversary's potential launch sites to improve the early-warning times further.

Conclusion

Although the operationalisation of hypersonic air power may provide the user with tactical advantages in the battlespace, contemporary air power will still be viable in a future battlespace. Thus, contemporary air power will remain relevant and effective, albeit with changed risks and new missions. The designs and applications for air power that persisted through the Cold War era under the threat of ICBMs boosting warhead to hypersonic speeds recognised the merit of addressing the risks posed by hypersonic ICBMs by considering the use of conventional air power in 'left-of-launch' and 'pre-emptive strike' missions. However, the continuing viability of contemporary air power employed during the Cold War was not made obsolete by ICBMs. Therefore, the risks from the operationalisation of new hypersonic manoeuvring air vehicles should be reviewed against the similar risks posed by ICBMs and known since the Cold War between the US and the Soviet Union.

Using the definition for disruption adopted in this chapter, hypersonic air power need not be considered a disruption to contemporary air power; instead, it is a disturbance that necessitates adaptation to contemporary air power's operational uses and risks. Hypersonic air power will not necessarily make non-hypersonic air power obsolete or irrelevant. The operationalisation of hypersonic propulsion need not be portrayed as a necessary change catalyst to invest in hypersonic capabilities and match strategic peers with hypersonic capabilities, nor consider contemporary air power obsolete. This view of the effect of hypersonic air power could be interpreted as being a quixotic moment.

Notes

1   Royal Australian Air Force, *Beyond the Planned Air Force: Thoughts on Future Drivers and Disruptors* (Canberra: Air and Space Power Centre, 2017), https://airpower.airforce.gov.au/publications/beyond-planned-air-force.

2 Department of Defence, *Integrated Investment Program* (Canberra: Commonwealth of Australia, 2016).

3 See, for example, Seng Boey Peter Dortmans and Joanne Nicholson, *Forward 2035 – DSTO Foresight Study* (Canberra: Defence Science & Technology Organisation, 2014); Paul Calhoun, "DARPA Emerging Technologies," *Strategic Studies Quarterly* 10, no. 3 (2016): 91–113; "Gartner's 2016 Hype Cycle for Emerging Technologies Identifies Three Key Trends That Organizations Must Track to Gain Competitive Advantage," *Newsroom, Gartner Inc.* Website, 16 August 2016, www.gartner.com/en/newsroom/press-releases/2016-08-16-gartners-2016-hype-cycle-for-emerging-technologies-identifies-three-key-trends-that-organizations-must-track-to-gain-competitive-advantage.

4 'Air power practitioner' is the collective term for those involved in the development and employment of air power. See Air and Space Power Centre, *The Air Power Manual*, 7th ed. (Canberra: Commonwealth of Australia, 22 March 2022), chap 1, https://airpower.airforce.gov.au/publications/APM7thEd.

5 See, for example, Travis Hallen and Michael Spencer, *Hypersonic Air Power* (Canberra: Air and Space Power Centre, 25 June 2018), https://airpower.airforce.gov.au/publications/hypersonic-air-power.

6 Brett Balinski, "Accelerating Hypersonic Research in Aust," *InnovationAus.com* Website, 6 July 2020, www.innovationaus.com/accelerating-hypersonic-research-in-aust/.

7 "Obituaries: Ray Stalker," *Australian Academy of Science* Website, March 2014, www.science.org.au/academy-newsletter/australian-academy-science-newsletter-95/obituaries-ray-stalker.

8 "HIFIRE Program," Defence Science Technology Group, Department of Defence Website, accessed 29 March 2022, www.dst.defence.gov.au/partnership/hifire-program.

9 Russell Boyce, "Opinion: Why the Scramjet Was Not a Failure," *Newsroom, UNSW Sydney* Website, 9 December 2013, https://newsroom.unsw.edu.au/news/science-technology/why-scramjet-was-not-failure.

10 Senator L. Reynolds, "Australia Collaborates with the US to Develop and Test High Speed Long-Range Hypersonic Weapons," *Department of Defence* Website, 1 December 2020, www.minister.defence.gov.au/minister/lreynolds/media-releases/australia-collaborates-us-develop-and-test-high-speed-long-range#:~:text=Last%20week%20Australia%20and%20the,size%20prototype%20hypersonic%20cruise%20missiles.

11 Senator M. Payne, "Hypersonic Flight Test a Success," *Department of Defence* Website, 10 July 2017, www.minister.defence.gov.au/minister/marise-payne/media-releases/hypersonic-flight-test-success.

12 Department of Defence, *2020 Defence Strategic Update* (Canberra: Commonwealth of Australia, 2020), 38.

13 See, for example, Richard H. Speier, et al., *Hypersonic Missile Nonproliferation* (Santa Monica: RAND Corporation, 2017).

14 Speier, et al., *Hypersonic Missile Nonproliferation,* 47.

15 Andrew Davies, "Hypersonic Weapons Are Coming – Whether We're Ready or Not," *The Strategist*, Australian Strategic Policy Institute Website, 26 March 2021, www.aspistrategist.org.au/hypersonic-weapons-are-coming-whether-were-ready-or-not/.

16 Lauren Caston, et al., *The Future of the U.S. Intercontinental Ballistic Missile Force* (Santa Monica: RAND Corporation, 2014), 33–34.

17 Michael Spencer, "Three Stages of the Inter-Continental Ballistic Missile (ICBM) Flight (#305)," *Pathfinder* 9 (2019), 85–92; "Ballistic Missile Defence System – A System of Elements," Missile Defense Agency, U.S. Department of Defense Website, updated 22 July 2021, https://mda.mil/system/elements.html.

18 Alan Weedon, "Chinese Navy Ship Seen Carrying an Apparent Railgun Capable of Firing Hypersonic Projectiles," *ABC News*, updated 3 January 2019, www.abc.net.au/news/2019-01-02/chinese-warship-with-electromagnetic-railguns-spotted-at-sea/10680108.

19 Franz-Stefan Gady, "China Tests New Weapon Capable of Breaching US Missile Defence Systems," *The Diplomat*, 28 April 2016, https://thediplomat.com/2016/04/china-tests-new-weapon-capable-of-breaching-u-s-missile-defense-systems.
20 "DF-ZF Hypersonic Glide Vehicle," Missile Defence Advocacy Alliance Website, March 2019, https://missiledefenseadvocacy.org/missile-threat-and-proliferation/todays-missile-threat/china/df-zf-hypersonic-glide-vehicle/.; Erika Solem and Karen Montague, "Updated–Chinese hypersonic weapons development," *The Jamestown Foundation*, 21 April 2016, https://jamestown.org/program/updated-chinese-hypersonic-weapons-development/.
21 Amelia Greig, "Fundamentals of Nuclear-Powered Engines," in *Nuclear Engine Air Power*, eds. David Burningham, Amelia Greig, Peter Layton, and Michael Spencer (Canberra: Air Power Development Centre, 2020), 15–44, https://airpower.airforce.gov.au/publications/nuclear-engine-air-power.
22 Fact sheet, National Aeronautics and Space Administration, "NASA Hyper-X Program Demonstrates Scramjet Technologies – X-43A Flight Makes Aviation History," NASAFacts, National Aeronautics and Space Administration Website, 2006, www.nasa.gov/centers/dryden/pdf/171371main_FS-040-DFRC.pdf.
23 "X-51A Waverider," United States Air Force Website, accessed 29 March 2021, www.af.mil/About-Us/Fact-Sheets/Display/Article/104467/x-51a-waverider/.
24 "SABRE – the Engine that Changes Everything," Reaction Engines Website, accessed 29 March 2021, https://reactionengines.co.uk/advanced-propulsion/sabre/#:~:text=of%20a%20rocket.-The%20engine%20that%20changes%20everything.,of%20sound%20for%20space%20access.
25 See Air and Space Power Centre, *The Air Power Manual*, chap 3.
26 Ibid., chap 2.
27 Calculated for a vehicle travelling at Mach 5 (6000 km/hr) over 1,000 km in about ten minutes.
28 See Air and Space Power Centre, *The Air Power Manual*, chap 4.

References

Government sources

Boey, Seng, Peter Dortmans, and Joanne Nicholson. *Forward 2035–DSTO Foresight Study*. Canberra: Defence Science & Technology Organisation, 2014.
Department of Defence. *Integrated Investment Program*. Canberra: Commonwealth of Australia, 2016. www.defence.gov.au/sites/default/files/2021-08/2016-Defence-Integrated-Investment-Program_0.pdf.
———. *2020 Defence Strategic Update*. Canberra: Commonwealth of Australia, 2020. www.defence.gov.au/about/publications/2020-defence-strategic-update.
Royal Australian Air Force. *Beyond the Planned Air Force: Thoughts on Future Drivers and Disruptors*. Canberra: Air and Space Power Centre, 2017.
———. *The Air Power Manual*. 7th ed. Canberra: Commonwealth of Australia, 2022. https://airpower.airforce.gov.au/publications/APM7thEd.

Published sources

Calhoun, Paul. "DARPA Emerging Technologies." *Strategic Studies Quarterly* 10, no. 3 (2016): 91–113.
Caston, Lauren, Robert S. Leonard, Christopher A. Mouton, Chad J. R. Ohlandt, S. Craig Moore, Raymond E. Conley, and Glenn Buchan. *The Future of the U.S. Intercontinental Ballistic Missile Force*. Santa Monica: RAND Corporation, 2014.

Greig, Amelia. "Fundamentals of Nuclear-Powered Engines." In *Nuclear Engine Air Power*, edited by David Burningham, Amelia Greig, Peter Layton, and Michael Spencer. Canberra: Air Power Development Centre, 2020.

Hallen, Travis, and Michael Spencer. *Hypersonic Air Power*. Canberra: Air and Space Power Centre, 25 June 2018.

Speier, Richard H., George Nacouzi, Carrie Lee, and Richard M. Moore. *Hypersonic Missile Nonproliferation*. Santa Monica: RAND Corporation, 2017.

11 Military culture and resistance to technical innovation

Matt Hegarty

In 2011, the United States Navy (USN) and Northrup Grumman began flight-testing the X-47B Uncrewed Combat Air Vehicle (UCAV), which had been purposely designed for aircraft carrier operations. Between the two test vehicles, they achieved remarkable results, including launches and recovery to aircraft carriers at sea. This experimental UCAV is especially notable because it was the first Uncrewed Aerial Vehicle (UAV) to be catapulted. So far, it is also the largest to operate autonomously from a US aircraft carrier. As part of the test programme, the X-47B successfully demonstrated autonomous launch and recovery, wave-offs (aborted landing requiring the vehicle to go around), touch and go landings to simulate a failure to catch the arrestor wire, and flying in the pattern with crewed aircraft. These achievements were impressive firsts, but perhaps most impressively, they demonstrated autonomous air-to-air refuelling behind an unmodified KC707 tanker. The USN's UCAV demonstrator programme has been, by all counts, very successful. Although the programme is now complete, and a follow-on uncrewed carrier-launched airborne surveillance and strike programme is underway, there is evidence of some resistance within the USN to embrace a true UCAV. Autonomous aircraft flight from an aircraft carrier is a complex operation, but UCAV operations in the US military are essentially 'mainstream'. The MQ-9 Reaper entered operational service in 2007, yet the US Navy does not have a UCAV equivalent in service.

The Australian experience is not too dissimilar. Despite a brief period of service with the medium-altitude long-endurance Heron, as of late 2021, the Australian Defence Force (ADF) still does not operate any UAVs other than small, tactically employed surveillance types.[1] However, there are some positive signs. The MQ-9B SkyGuardian and the MQ-4C Triton are scheduled to enter RAAF service in the mid-2020s. Meanwhile, the partnership between Boeing and the RAAF has resulted in several successful test flights of a prototype UCAV, originally named 'Loyal Wingman' and subsequently designated as the MQ-28A 'Ghost Bat'. The increasing use of automation and the delivery of uncrewed aviation platforms is beyond doubt. What is yet to be demonstrated is how well the existing Air Force culture is ready to adopt these new systems and make the best of the military advantages they offer.[2]

DOI: 10.4324/9781003230656-15

In *The National Interest*, James Holmes noted two opposing theories on technological progress. On the one hand, he referenced philosopher Karl Popper, who stated that a hypothesis is developed to explain or predict reality and then is tested to try to 'falsify' or disprove it through rigorous experimentation. Popper's theory is principally a scientific notion. On the other, Thomas Kuhn argued that scientific progress occurs through a fractious political process whereby the existing doctrine and culture have gatekeepers with vested interests in preserving the status quo.[3] This theory suggests technological progress is socially governed. That there would be resistance to the increasing introduction of uncrewed aircraft into military service appears reasonably obvious. Culturally, we appear pre-disposed to the romance of the gentlemanly duel between warriors who are reliant on their raw skills to fly and fight. Popular science-fiction continues to feature pilots at the controls of spacecraft, operating flight controls and aiming weapons systems as if flying a Supermarine Spitfire against Messerschmitt Bf109 in the Battle of Britain, despite the increasing sophistication of modern aviation systems and the accompanying autonomy such systems require.

So far, there is little evidence that UAVs or UCAVs have displaced pilots, but this is only a matter of time. However, the answer to why pilots would reject the notion of uncrewed vehicles is not just about the loss of flying jobs—there is a more complex set of issues at play. This chapter discusses the relationship between organisational culture and technological innovation, using the development of Blitzkrieg as a historical example. The focus on culture is important because it affects how the Royal Australian Air Force (RAAF), and the ADF more broadly, embraces technological innovation, or indeed, whether it embraces it at all.

The Fourth Industrial Revolution

In 2016, Klaus Schwab, founder and executive chairman of the World Economic Forum, famously defined the Fourth Industrial Revolution as 'the advent of "cyber-physical systems" involving new capabilities for people and machines'. He notes:

> The Fourth Industrial Revolution represents entirely new ways in which technology becomes embedded within societies and even human bodies. Examples include genome editing, new forms of machine intelligence, breakthrough materials and approaches to governance that rely on cryptographic methods such as the blockchain.[4]

There are many consequences of this revolution. The ADF will not only need to remain savvy regarding the opportunities that these new technologies will present; it will also need to remain alert to the fact that such technologies are becoming increasingly prolific. Much of the technology will be within the means of all except the poorest of nations, and our reliance on holding a technological edge in the face of numerically superior regional forces will eventually fail to provide us with the advantages we seek. That is why we must know and understand our military culture and the sub-cultures. Emerging technology and the innovation that combine to bring us new capabilities and military advantage will only succeed if the ADF remains open to change and alert to its biases.

Edgar Schein defines organisational culture as:

> A pattern of basic assumptions—invented, discovered, or developed by a given group as it learns to cope with its problems of external adaptation and internal integration—that has worked well enough to be considered valid and, therefore, to be taught to new members as the correct way to perceive, think, and feel in relation to these problems.[5]

Our organisational (cultural) inertia, or the resistance to change that lies within the military's warfighting concepts, can help prevent the whimsical pursuit of technology for the sake of technology itself. However, equally, it can stymie the progress of warfighting capability and leave one military force at a significant disadvantage in the face of another which has embraced new technology and developed supporting doctrine to maximise the effect of its employment. One example of the effect that technological development and supporting doctrine can have in the application of military power is the development of Blitzkrieg.

A historical case study: the origins of blitzkrieg

Blitzkrieg is sometimes described as a land warfare concept first introduced by Nazi Germany when its army first invaded Poland in 1939 and then very successfully in the subsequent invasion of France and the Low Countries in 1940. However, its origins lie neither in the Second World War nor the Spanish Civil War during the late 1930s, during which technological aspects were tested. Rather, the origins of Blitzkrieg are found in the First World War. Arguably, these origins lie with the British Army.

Looking at the conditions underpinning the way of war on the Western Front in the First World War is instructive. The Western Front was defined by trench warfare, stalemate, and mass slaughter. Three technological innovations irrevocably changed the land battle: machine guns, the increased rates of fire and accuracy of artillery, and aircraft. The open field manoeuvres for which commanders on both sides had planned in the lead-up to the war had become impossible. Trenches, barbed wire, and new weaponry now favoured defence over attack. When significant breaches in the front were achieved, they were often plugged with reinforcements supported by artillery while attacking forces were estranged from their own. Static machine guns in defensive positions more than made up for the manoeuvrability of infantry carrying bolt action rifles.

In 1915, efforts were underway to build a new machine that would combine manoeuvrability, firepower, and protection, and thus, the British Mk1 tank was born. First used with mixed success in September 1916, the Mk1 male (there were two versions) weighed nearly 30 imperial tons, was powered by a 78 kW engine, had a range of 37 kilometres at 4.5 kilometres per hour, and was armed with two six-pound guns and three machine guns. Proponents of the tank included the first commander of the Tank Corps, Brigadier General Hugh Elles, and his Chief of Staff, Lieutenant Colonel John Frederick Charles (J.F.C.) Fuller, who was later promoted to major general and became a noted strategist and military historian. However, their debut on the battlefield was marred by small numbers and reliability problems. They were used again in the Battle of Passchendaele in Autumn 1917, in unusually wet conditions, and struggled with the mud, but in late 1917, Elles pressed for their massed use in the Battle of Cambrai.[6] Used in this way, Elles achieved considerable success, but it was not a simple matter of superior technology. How the technology was employed was an essential ingredient. Strategist and historian Basil Liddell-Hart later referred to the method used as the 'indirect approach', and this combination of technology and technique eventually became known as Blitzkrieg.

In describing the advent and development of blitzkrieg, Perrett notes:

> The basic principle of Blitzkrieg is to seek out for attack the points where the enemy is weakest and least expecting to be attacked; then having broken in to secure a foothold, to pour in whatever can be found to develop out of the break-in, a breakthrough.[7]

The use of massed armour in this way is now accepted as standard operational doctrine, but in the First World War, both the technology and the technique for employment attracted considerable resistance. Despite the slaughter and stalemate that had dominated through years of war already, Field Marshal Douglas Haig, who commanded the British Expeditionary Force from late 1915 until the end of the war, willingly endorsed the obscene carnage of the battles of the Somme in 1916 and Passchendaele in 1917. Like many planners of that time, he continued to believe that the big breakthrough would eventually come. Attitudes continued to hold that the new technologies like aircraft and tanks would be little more than a distraction to the main effort.[8] While there were usually some changes to the configuration in each new assault, they tended to be minor variations or simply a matter of increasing the scale of an assault (e.g., an attack across a wider front or an artillery bombardment that lasted longer before the infantry assault). Throughout this period, the prevailing culture held the superiority of infantry as the main weapon.

Eventually, the use of armour came to be accepted—perhaps not universally, but to the point where it was used decisively in the summer and autumn campaigns of 1918. Indeed, one of the most famous battles, the Battle of Hamel, which involved Australians and New Zealanders fighting for the first time alongside the US Army, included the use of British tanks and aircraft. By the end of the war, Britain and France had fielded thousands of tanks between them, contributing to the Allied victory. Yet despite having witnessed their effective use directly, Germany fielded very few—the Germans produced only 20 of their A7V tanks, and they occasionally included captured British or French tanks in their attacks.[9] The initial successes of the German stormtrooper tactics saw elite infantry formations rushing forward fast and deep into enemy territory with close support from artillery and armed aircraft. These tactics adhered to the same principles of indirect attack, as did the British use of armour, but there was no widespread acceptance or urgency to adopt the tank.

Despite having pioneered the initial successful use of massed armour 20 years earlier, British and French forces were not able to reproduce their own versions of Blitzkrieg in 1940, much less defend against them. There are likely many reasons for the failure of the French and British forces to embrace armour, including overconfidence in defensive systems such as the Maginot Line, but there was also an institutional resistance. In 1919, in a stunning lack of imagination and forethought, Major General Sir Louis Jackson of the British Army stated during an address to the Royal United Services Institute:

> The tank proper was a freak. . . . The circumstances which called it into existence were exceptional and not likely to occur again. If they do, they can be dealt with by other means.[10]

Indeed, Britain neglected the ongoing development of the tank after the First World War, preferring instead to believe that the infantry and cavalry would remain the primary battlefield combat branches. Perrett notes that '[t]he Twenties has been quoted as the decade in which the British Army obstinately continued to pay more for its fodder than for its fuel'.[11] Having pioneered the use of armour and the tank during the First World War, British incorporation of armour between the wars was subsequently half-hearted.

In contrast, Nazi Germany, having little more than dallied with armour during the First World War, subsequently saw the potential of the capability and made armour the centrepiece of its attack on Europe and North Africa. General Heinz Guderian, known as the architect of the *Panzerwaffe*, led the German Panzer Corps in the invasion of Poland. In an interesting twist, Guderian credited the writings of J.F.C. Fuller and Basil Liddell-Hart as the main inspiration for his leadership in developing Blitzkrieg as a technique.[12] It is worth pondering the cultures of the respective armies and how these cultures embraced or struggled with the incorporation of armour and its advantages.

The stilted acceptance and incorporation of armour in the early 20th century is not an unfamiliar story. Similar stories abound regarding the initial use and incorporation of aircraft, which were initially seen as little more than surveillance platforms. In the maritime environment, the larger navies initially struggled to accept that the battleship would cease to be the main strike platform as aircraft carriers grew in both size and number. Now, there is a growing debate surrounding whether the aircraft carrier can remain the principal maritime strike platform given weapons being developed under the banner of anti-access and area denial, such as the hypersonic glide weapon. The X-47B and its successor, uncrewed carrier-launched airborne surveillance and strike (UCLASS), aimed to reassert the superiority of American aircraft carriers. However, success will depend on the willingness of the USN to embrace the opportunities that these new technologies offer. Notably, the UCLASS has since been cancelled, and, in its place, a new programme called Carrier-Based Aerial Refuelling System has commenced. Uncrewed strike has been deferred to a future programme.[13] Kuhn's 'fractious political process' may well be at play, given the proximity of the USN programme to field a sixth-generation fighter ('F/A-XX') to replace the F/A-18 E/F Super Hornets.[14]

The present day

In considering emerging technologies and their influence on air and space power, a retrospective look at the introduction of the tank and the changes to warfare that it brought provides a worthy lesson. Just as worthy is an understanding of the institutional biases which affected militaries in 1918 and an awareness that some continue to affect us now. To some degree, the traits that create the conditions necessary to make good soldiers, sailors, aviators, and officers often conspire to slow the adoption of change, especially when their identity appears to be threatened. Writing in the US Army Journal *Parameters*, Andrew Hill observes that the military is a very significant and somewhat conservative bureaucracy and offers an explanation as to why innovation faces opposition from within:

> Militaries are bureaucracies that depend on standardisation of tools, training, methods, and organization. Innovation subverts this standardization and consistency, first, in the exploration of a new approach . . . and then (if the innovation is successful enough) in the eventual replacement of the existing approach throughout the organization.[15]

Notably, the eventual incorporation of the innovation, if it is to occur, requires organisational change. This requires a change to the military organisation's culture for the larger innovations.

Elting Morison writes about the instinct to protect oneself and observes a plain truth when he states that military organisations are built around weapon systems:

> The opposition, where it occurs, of the soldier and sailor to [innovation] springs from the normal human instinct to protect oneself, and, more especially, one's way of life. Military organizations are societies built around and upon the prevailing weapons systems. Intuitively and quite correctly the military man feels that a change in weapons portends a change in the arrangements of his society.[16]

Hence, the RAAF should expect no less than a defensive posture from those groups most threatened by the proposal to introduce new technology that lessens the role of a specific weapon system—or worse, removes the human from that role entirely.

Conclusion

Having discussed a worthy historical example of the impact of military culture on the adoption of emerging technology and noting the problem persists, we might ask, so what? The answer: the ADF appears to be experiencing the most fundamental change to its tools of war in generations. This change is driven not by the Defence industry's pursuit of military technology but principally by the innovation and adoption of technology essential to the progeny of the industrial and commercial realm. This drive for innovation is not a passing fad; it is a new industrial revolution. Australia's Defence community, and the RAAF in particular, would do well to ponder the consequences of constructing a binary choice between Popper's conceptualisation of progress as generally scientifically-led and Kuhn's view that emphasises the socially constructed nature of technological development. Adherents to the former demand empirical proof indicating the superiority of a proposed innovation, a particularly difficult task, given the pace of development and the relatively few opportunities to field high-end warfighting capabilities outside of conflict. Conversely, social constructivists emphasise the effect of gatekeepers in impeding scientific 'revolution' if a proposed technology contrasts with the accepted, hegemonic military identity.

Fortunately, a binary choice is not necessary. It is possible to value scientific rigour while also appreciating and managing the 'fractious political processes' that threaten the take-up of new technology. Such an approach is articulated in the RAAF's *Air Force Strategy*.[17] The strategy encourages scientific rigour in technological development by placing emphasis on 'provid[ing] the opportunity and space for members to curate leading-edge research, honing ideas through to realisation in a safe environment that is failure tolerant'.[18] This direction is complemented by measures aimed at diluting the power of hegemonic groups to act as gatekeepers of vested interests by ensuring 'the Air Force SLT [Senior Leadership Team] contains a broad mix of the best leaders and thinkers in the organisation, irrespective of their tactical and technical background'.[19] The case study outlined in this chapter attests to the challenges facing those who seek to change deep-seated military cultures, so the RAAF would do well to ensure it has a robust implementation plan to ensure its strategy represents more than empty platitudes.

Notes

1 The Heron was operated by the RAAF from 2009 to 2017. "RAAF Retires 'Legacy' Heron UAV," *Australian Aviation* Website, 9 August 2017, https://australianaviation.com.au/2017/08/raaf-retires-legacy-heron/. The Australian Army and Royal Australian Navy continue to operate various tactical UAVs including the RQ-7B Shadow 200 and S-100 Camcopter.

2 Nigel Pittaway, "$200m Thrust in Northern Territory for Aircraft Base to Service Triton," *The Australian,* 30 October 2021; "AIR7003 Phase1MQ-9B SkyGuardian Armed Remotely Piloted Aircraft System," Department of Defence Website, accessed 15 November 2021, www.defence.gov.au/project/air7003-skyguardian-armed-remotely-piloted-aircraft-system; Inder Singh Bisht, "Boeing Australia Tests Unmanned 'Loyal Wingman' Aircraft," *The Defense Post,* 5 November 2021, www.thedefensepost.com/2021/11/05/boeing-australia-loyal-wingman/.

3 James Holmes, "The Mighty X47B: Is It Really Time for Retirement?" *The National Interest*, 6 May 2015, https://nationalinterest.org/feature/the-mighty-x-47b-it-really-time-retirement-12818.
4 Nicholas Davis, "What Is the Fourth Industrial Revolution?" *World Economic Forum Website*, 19 January 2016, www.weforum.org/agenda/2016/01/what-is-the-fourth-industrial-revolution/.
5 Edgar Schein, *Organizational Culture and Leadership* (New York: Jossey Bass, 2010), 17.
6 For more on the effects of mud at the Battle of Passchendaele, see: C. E. Wood, *Mud: A Military History* (Washington, DC: Potomac Books, 2006), 23–24, ProQuest. For more on the massed use of tanks, see: Kristafer Ailslieger, "The Battle of Cambrai," *Armor* 109, no. 5 (September/October 2000): 34–36.
7 Bryan Perrett, *A History of Blitzkrieg* (London: Robert Hale Ltd, 1983), 12.
8 Ibid., 12.
9 "Early Tank Design and Use," Australian War Memorial Website, last modified 14 April 2020, www.awm.gov.au/learn/schools/resources/mephisto/design-use.
10 Perrett, *A History of Blitzkrieg,* 54.
11 Ibid., 55.
12 Heinz Guderian, *Panzer Leader* (London: Michael Joseph, 1970), 20.
13 Mackenzie Eaglen and Bryan McGrath, "Future Warfare: X-47B," Real Clear Defense Website, 24 April 2015, www.realcleardefense.com/articles/2015/04/25/future_warfare_the_x-47b_107890.html.
14 Inder Singh Bisht, "US Navy to Replace Super Hornets with Mix of Manned, Unmanned Aircraft," *The Defense Post,* 1 April 2021, www.thedefensepost.com/2021/04/01/us-navy-to-replace-super-hornets/.
15 Andrew Hill, "Military Innovation and Military Culture," *Parameters* 45, no. 1 (Spring 2015), 85.
16 Elting E. Morison, "A Case Study of Innovation," *Engineering and Science Monthly* 13, no. 7 (April 1950), 8.
17 Royal Australian Air Force, *Air Force Strategy: Key Highlights* (Canberra: Commonwealth of Australia, 2020), 14.
18 Ibid., 14.
19 Ibid., 13.

References

Government sources

Royal Australian Air Force. *Air Force Strategy: Key Highlights*. Canberra: Commonwealth of Australia, 2020. www.airforce.gov.au/sites/default/files/air_force_strategy.pdf.

Published sources

Ailslieger, Kristafer. "The Battle of Cambrai." *Armor* 109, no. 5 (September/October 2000): 34–36.
Guderian, Heinz. *Panzer Leader*. London: Michael Joseph, 1970.
Hill, Andrew. "Military Innovation and Military Culture." *Parameters* 45, no. 1 (Spring 2015): 85–98.
Morison, Elting E. "A Case Study of Innovation." *Engineering and Science Monthly* 13, no. 7 (April 1950): 5–11.
Perrett, Bryan. *A History of Blitzkrieg*. London: Robert Hale Ltd, 1983.
Schein, Edgar. *Organizational Culture and Leadership*. New York: Jossey Bass, 2010.
Wood, C. E., *Mud: A Military History*. Washington, DC: Potomac Books, 2006. ProQuest.

12 Rubbery assumptions

Anti-G suits and air power in the Second World War

Peter Hobbins

'At first I was a little suspicious of the Anti "G" Suit and entered all manoeuvres somewhat gingerly', reported Squadron Leader Sidney 'Beau' Carr, 'but [I] soon acquired the utmost confidence'. Indeed, he added,

> I consider the Cotton Anti 'G' Suit to be among the most notable advances in aviation that the war has produced, and the advantages that a fighter pilot wearing the suit would have over any opponent not so equipped are enormous.[1]

Authored in June 1943, Carr's report was circulated at a critical moment for Allied air power projections. A highly experienced Royal Air Force (RAF) pilot, Carr had been sent to Australia in a technical liaison capacity. His visit coincided with the first field trials of the Cotton Aerodynamic Anti-G (CAAG) suit, then being conducted in Darwin by No. 452 Squadron Royal Australian Air Force (RAAF).[2] This local innovation had been spearheaded by The University of Sydney medical scientist Frank Cotton, who was inspired by reports of massed dogfights during the Battle of Britain in 1940.[3]

Carr was no stranger to Australia nor the RAAF. Before the war, he had been posted to the Service's primary training station at Point Cook, where he served as chief flying instructor until he returned to Britain in 1941. Shortly before departing, Carr met with Cotton, who, 'like all boffins, had tremendous faith in his own theories'.[4] From September 1940, the theory that Cotton pursued with unabated vigour centred around 'a promising approach to the problem of "Black-out" in pilots in fighter planes'.[5] Few denied its potential; within months, he had secured financial, administrative and logistical support from The University of Sydney, the new National Health and Medical Research Council (NHMRC), and the RAAF.

Cotton's theory was simple in concept and unique in execution. A talented athlete and a specialist in the human cardiovascular system, he carefully considered the problem of G-induced loss of consciousness. When subjected to raised centrifugal forces during high-speed turns or pull-out from steep dives, aviators progressively lose their vision before losing consciousness. These phenomena are caused by blood draining from the head and chest under high-G conditions, subsequently pooling in the abdomen and legs. Starving the retinas of blood leads first

DOI: 10.4324/9781003230656-16

to grey-out and then total black-out of sight, while lack of blood flow to the brain causes a rapid loss of consciousness. All bode ill for the control of an aircraft during extreme manoeuvres—especially in combat.

In a classic 'Eureka' moment, Cotton envisioned a solution. He proposed a rubberised suit that encased flyers and inflated rapidly under elevated G-force. The suit was designed around his 'gradient pressure principle', meaning that instead of a uniform inflation, the outfit squeezed the aviator's body most tightly in the feet, with additional compartments providing lesser levels of pressure up to the bottom of the rib cage. This 'pneumodynamic' response meant that the greatest compression was applied where the most potent effects of high G manifested—the lower legs. The suit operated via compressed gas, with the pressure compartments differentially inflated via a series of graduated valves.

As I have outlined in previous publications, Cotton's programme rapidly spiralled in complexity.[6] The CAAG suit itself went through multiple prototypes and variants, ultimately shedding the troublesome inflatable boots that interfered with rudder control and hence gunnery. The novel latex construction frequently ruptured, while robust zippers essential to rapid donning and doffing proved almost impossible to import owing to wartime scarcity. Few Australian firms could manufacture the gradient-pressure valves to the required tolerances, while onboard compressors gave way to bottled gases. A warning light and buzzer system was required to alert G-protected pilots that they might overstress their aircraft during high-energy manoeuvres, while another circuit warned when turns became so tight that the aircraft was at risk of stalling. Finally, the human centrifuge built at The University of Sydney with NHMRC and RAAF funding regularly broke down, compromising experimental work, prototype testing, and 'indoctrination' of Air Force flyers into the use of the CAAG suit.

And yet, the device worked. Experimental studies in the centrifuge—using Cotton as the 'guinea pig'—confirmed that when the CAAG suit inflated as planned, it increased tolerance to centrifugal forces by around 2 G. Translated into air combat terms, this might either allow a tighter manoeuvre—such as turning inside an enemy aircraft—or maintaining consciousness for several critical seconds while rapidly pulling out of a dive. Test flights in a Commonwealth Aircraft Corporation (CAC) Wirraway and then the RAAF's sole Hawker Hurricane led service pilots to praise Cotton's invention. By late 1942, advanced design was underway to install CAAG equipment into the RAAF's frontline CAC Boomerangs and Curtiss P-40 Kittyhawks, although ultimately, it only became operational in Supermarine Spitfires. Tests conducted in Brisbane in August 1943 against a captured Mitsubishi A6M3 Zero 'Hap' were shared via a RAAF technical bulletin. It noted that a 'special' Spitfire Mk V outfitted with CAAG apparatus 'was able to evade and out-manoeuvre [the] Hap by combining high speed and High "G", but lost this advantage when airspeeds fell below 250 miles [400 kilometres] per hour'.[7] After three years of trials—both experimental and logistical—Cotton's theory, at last, appeared to be vindicated.

It was at this moment that Carr returned to Australia, and No. 452 Squadron evaluated the Cotton suit under frontline conditions. From June to November 1943, their Spitfire Mk Vs were converted to deploy this top-secret equipment. 'The increased manoeuvrability at high speeds should prove a great asset in shooting and taking evasive action from the Zero fighter', enthused Flying Officer Johnny Bisley.[8] However, it appears that the only occasion on which the outfit was actually tested in action occurred during the interception of a Japanese reconnaissance in force over Darwin on 7 September. Few conclusions were drawn from this encounter, other than the propensity for Spitfire pilots wearing the Cotton suit to overstress their aircraft in the heat of dogfighting.[9]

While many of Bisley's comrades agreed with his assessment of the suit's potential, they also noted profound disadvantages. These included its cumbersome bulk and weight (nine kilograms), plus the unbearable heat and discomfort of being encased in a tightly laced rubber costume in Australia's Top End, where temperatures regularly surpass 30 degrees Celsius. Despite the provision of an air-conditioned chocolate-delivery van to act as a temporary ready room, the receding Japanese threat saw the CAAG suits withdrawn from operations on 9 November.

A merry-go-round of technical exchange

Yet those five months in Darwin came at a crucial time for the western Allied air forces. This period encompassed the conclusion of the African campaign, followed by the invasions of Sicily and Italy. Preparations were well underway for the forthcoming invasion of western Europe, including the necessity for air superiority across the continent and air supremacy above the landing beaches. While the Japanese were on the defensive in the southwest Pacific, their nimble army and navy fighters continued to imperil American and Australian bombing campaigns and amphibious assaults. Meanwhile, in the Burma theatre, the impending Allied push eastwards from India faced air parity against the well-equipped Imperial Japanese Army Air Force.[10] Therefore, tactical operations would benefit from a rapidly deployed device that might wrest a decisive air combat advantage on all fronts.

At this juncture—late in 1943—Squadron Leader Carr returned to Britain. He was posted to Farnborough specifically to oversee the anti-G programme being conducted by the Physiological Laboratory Royal of the Aircraft Establishment. Affirming his belief in the potential combat value of the Australian suit, Carr conveyed two CAAG outfits and their associated fittings home with him. While at Farnborough, his primary focus was to compare Cotton's garment against its British equivalent, the Franks Flying Suit (FFS), developed by a Canadian medical scientist, Wilbur 'Bill' Franks.

The gestation of the FFS had been even more protracted than its antipodean counterpart. Since 1938, Franks had pursued a completely different line of anti-G protection. Although his outfit also consisted of a rubberised outfit that embraced the legs and abdomen, it was filled with water rather than gas. The logic behind this 'hydrodynamic' mechanism was that as centrifugal forces drained blood towards the lower body, the water in the suit would also flow in that direction and provide a compensatory compression. Tested in a human centrifuge built in

Ottawa with Royal Canadian Air Force (RCAF) funding, the Franks suit offered 2 G of protection beyond the average pilot's physiological limit.

From 1940, the development of the FFS was supported by the RAF. In November 1942, it had been operationally tested by Fleet Air Arm (FAA) pilots during Operation Torch over North Africa.[11] Since then, however, the suit had been held in operational reserve, despite the Admiralty's keenness to deploy it during the Italian campaigns. Fear of exposing its secrets to the enemy meant that thousands of Franks suits were securely warehoused in anticipation of providing a surprise tactical advantage during Operation Overlord—the invasion of France.[12]

Carr's intelligence about the Australian anti-G suit was not unprecedented. In late 1941 and early 1942, Australian medical scientist Charles Kellaway— Director of Melbourne's Walter and Eliza Hall Institute—undertook a technical tour of inspection across North America and Britain. He reported back to the RAAF and NHMRC on numerous advances in Allied aviation equipment, ranging from oxygen masks to instrument panel lighting.[13] Before his departure in September 1941, Kellaway had written to Cotton requesting 'a few notes about the principle of your anti-g method, as these would be very useful to me in discussing the matter with the British authorities'.[14]

Cotton learned about the hitherto top-secret British anti-G programme during the subsequent imperial exchanges. He was soon authorised to travel in Kellaway's wake, arriving in the United States (US) just days before the Japanese attacks on Hawaii and Malaya in December 1941. Over the following months, Cotton discussed his nascent device with US Navy (USN), US Army Air Corps, and RCAF medical officers before crossing the Atlantic Ocean to meet with Franks in early 1942. Although the Canadian scientist at first attempted to incorporate Cotton's distinct principle into a generic Commonwealth patent for anti-G suits, the pair ultimately agreed to sustain two discrete lines of development. The RAF and its Flying Personnel Research Committee (FPRC) certainly encouraged Cotton and the RAAF to maintain the Australian project. British planners saw the CAAG suit as a backup to their more advanced FFS programme already in production at the Dunlop factory in Manchester.

Wartime technical exchanges remain poorly served by the historiography, apart from selected innovations such as radar and intelligence about enemy aircraft.[15] Yet, by 1942, the circulation of aviation medicine research entailed a vast and global enterprise. From 1943, the escalation of Allied air transport capability—and the diminishing threat of interception—led to a veritable merry-go-round of experts, documents, devices, and data that ranged from secret laboratories to participation in operational sorties.[16] Usefully for historians, the often-frank critiques by Allied visitors to Australian facilities provided a potent counterpoint to local hubris.[17] Australian researchers, conversely, enjoyed access to overseas facilities and information via scientific liaison offices in London and Washington, DC. Funded by the RAAF, the Flying Personnel Research Units at The Universities of Melbourne and Sydney also benefited from a welter of microfilmed American, Canadian, British and (occasionally) Soviet aviation medicine reports.[18]

Encapsulating the intense pace of this research, the RAAF's own FPRC was established in 1940 and published 140 reports and minutes by the time Japan surrendered. At least 48 of these documents—comprising more than one-third of the committee's entire wartime output—included references to anti-G work.[19] The British FPRC likewise released at least 630 reports between its establishment in September 1939 and August 1945.[20] Issued in February 1944, Volume No. 567 comprised Carr's comparative evaluation of the Cotton and Franks suits. His tone was now markedly less optimistic than the heady assessment that opened this chapter. Carr opined that few pilots were willing to wear either the Franks or Cotton suits, 'with their attendant discomforts of restriction, sweating and general sensation of being trussed up'.[21] Instead, he suggested, a similarly protective effect against high G could be attained by changes in aircraft design—notably a reclined pilot's seat and raised rudder pedals—plus the simple expedient of tensing the stomach muscles while manoeuvring.

A massive lacuna in air power planning

Given the pressing tactical need for an advantage over the latest German and Japanese aircraft, what had caused such a rapid turnaround in Carr's assessment of the empire's anti-G suits? Three key factors were at play: poor planning, American competition, and aircrew resistance. Together, these factors led to the effective abandonment of anti-G devices and their purportedly 'decisive' air power advantage—at least within Commonwealth air forces.

Coinciding with the RAAF's evaluation of the CAAG suit in Darwin, during mid-to-late 1943, the RAF also issued the Franks suit for overseas trials. Bearing in mind that avoiding enemy capture of the garment was an absolute necessity, the chosen locations were both islands: Malta and Ceylon (now Sri Lanka). By the time No. 145 Squadron was operational with the FFS in Malta, Axis forces had been evicted from North Africa, and enemy air activity over the island effectively ceased. In Ceylon, however, No. 17 Squadron RAF was commanded by Squadron Leader Montague 'Monty' Cotton—an Australian who happened to be the cousin of Frank Cotton. From November 1943 to January 1944, his Hurricane-equipped unit evaluated the FFS for defensive sorties under tropical conditions.[22]

In line with No. 452 Squadron's assessment of the Australian garment, RAF pilots were soon convinced that the Canadian anti-G suit would offer a distinct edge against Japanese fighters. For instance, Canadian-born Warrant Officer Harold Dow participated in a mock dogfight with Monty Cotton. While both were flying tropicalised Hurricane Mk IIs, only Cotton was donning an FFS. At airspeeds above 200 miles (320 kilometres) per hour, Dow proposed, 'without the suit it would be almost impossible to shoot down anyone wearing it'. Furthermore, while his commanding officer remained fresh after their vigorous aerobatics, Dow 'was completed fagged out and had to go to bed immediately after dinner'.[23] This report affirmed another benefit of the Franks suit: a pronounced reduction in fatigue after strenuous manoeuvring. This improvement in personal efficiency was widely noted by air force and Admiralty aviators.

However, despite his considerable enthusiasm for the FFS, Monty Cotton acknowledged that the rubberised suit was completely unsuited to the local climate. 'The question of overheating on the ground is our worse [sic] problem', he stated unequivocally, 'and under present conditions seems almost insurmountable'.[24] But the suit was also unwelcome in temperate environments. After trials undertaken in Spitfire Mk Vs over Scotland, the RAF's No. 485 (New Zealand) Squadron rejected the Franks suit because it considerably diminished the ability of pilots to turn their heads and watch for enemy aircraft behind them. Receipt of these negative assessments in January 1944 likely played a part in Carr's lacklustre evaluation of both the Australian and Canadian anti-G outfits a month later.

Nevertheless, by early 1944, RAF planners had surmised that the Franks suit was 'likely to be of much more value to pilots fighting the manoeuverable [sic] Japanese aircraft than to pilots on the Western front who already have the advantage in manoeuverability [sic] over their German opposite numbers'.[25] Admiralty carrier pilots remained enthusiastic about deploying the FFS in any theatre—especially among units flying Grumman Martlets and F6F Hellcats, which featured much roomier cockpits than Supermarine Seafires. On 26 February, the RAF finally decreed that all Allied anti-G suits would be downgraded to secret and released for operations over enemy territory from 1 April.

Here a massive lacuna appeared in British air power planning. 'This problem is rather like that of the drop tank', noted the RAF's Directorate of Operations, meaning that, at last, the tactical advantage was seen to outweigh the risk of exposing a new technology to the enemy. 'The Frank [*sic*] Suit is for the offensive . . . a decision cannot be reached unless the suit is tried out in actual operations'.[26] The analogy was imperfect, however. Drop tanks were another innovation that the Admiralty had repeatedly championed against air force conservatism.[27] Yet their value in extending the range of fighter aircraft was unequivocal, whether deployed operationally or otherwise. In contrast, the employment of anti-G suits could only be optimised once the appropriate tactics, manoeuvres, and training were developed to combat the enemy's latest frontline fighters. The jet-powered Messerschmitt Me 262, for instance, commenced service trials in April 1944 and became fully operational that July.

Astoundingly, the RAF headed towards D-Day with thousands of Franks suits stockpiled under armed guard but lacking any formal tactical preparedness. Apart from a repeatedly delayed instructional film, there were no manuals, lessons, demonstrations, or experienced fighter pilots to champion the rapid adoption of the FFS at the squadron level.[28] Yet since 1941, the Franks suit had been evaluated by the RAF's Air Fighting Development Unit and the Admiralty's equivalent Naval Air Fighting Development Unit. Both drew lessons from the experiences of Nos. 801 and 807 Squadrons FAA, whose Seafire pilots had worn the anti-G garment when fighting Vichy French Dewoitine D.520s during Operation Torch. Little effort, it seems, had since been devoted to operationalising the anti-blackout advantages offered by this novel garment. The only unit to have comprehensively assessed, documented, and disseminated air combat tactics for the FFS was Monty Cotton's No. 17 Squadron—on the Burma–India front.[29]

There had been no anti-G evolution in fighter-versus-fighter tactics in the European theatre, nor was it established whether the Franks suit improved the safety or effectiveness of dive-bombing and strafing. With Overlord just months away, it was almost too late. Belated reports from operations over France in April–May 1944 saw the FFS damned by Spitfire Mk IX pilots in the RAF's No. 66 Squadron, plus Nos. 331 and 332 (Norwegian) Squadrons. With no time left for modifications to the garment, and no opportunity to devise, promulgate, and implement training to capitalise on its benefits, the FFS was withdrawn from RAF use just weeks before the invasion of Europe.

Indoctrination and emasculation

The belated British decision to release anti-G gear for offensive operations, however, had finally freed the RAAF to deploy the CAAG suit overseas. From 17 April 1944, preparations began to fully outfit three Spitfire Mk VIII units with modified CAAG suits. By late July, No. 452 Squadron and two Australian-based RAF squadrons—Nos. 548 and 549 Squadrons—were fully kitted out with anti-G gear. This decision was likely driven not only by the shift in British policy but also by positive reports received from USN Hellcat pilots deploying their own gas-operated G-1 suits over Palau in March.[30] At the same time, the US Army Air Forces (USAAF) in Europe were also issuing large numbers of their air-operated G-2 and G-3 anti-G suits to Republic P-47 Thunderbolt and North American P-51 Mustang units as the invasion of Europe progressed.[31]

Here, again, the Allies' paths digressed. Typically, the USN and USAAF had developed their own discrete anti-G programmes, albeit with considerable consultation and crossover. Both Services had proved highly receptive to Cotton's 'gradient pressure principle' and, in fact, his theory dominated US developments from late 1941 into 1943.[32] The USAAF in Europe also evaluated the Franks suit—at one point ordering 500 from Britain—but then abandoned it at the same time as the RAF. However, as the American aeromedical establishment grew, its resources, experimental projects, and service integration did too.[33] By 1943, flight surgeons and medical scientists regularly ranged across experimental laboratories such as the Mayo Clinic in New York state, major domestic air bases such as Wright Field in Ohio, and 8th and 9th USAAF units based in Britain.

Soon the army and navy programmes converged on a common point: simplicity. American flyers demanded suits that were light, easy to wear, simple to operate and created minimal restriction of movement—including emergency egress from their aircraft. By early 1944, these imperatives led to the abandonment of the gradient pressure principle, with flight surgeons acknowledging that suits that inflated at a single pressure were far simpler to produce and offered only a slight decrement in anti-G protection. The telling point was that, unlike the 'insurmountable' discomfort that dogged both RAF and RAAF anti-G equipment, by D-Day, US aviators in Europe enjoyed comfortable, lightweight suits that lessened fatigue on long escort missions and offered air combat options for emerging threats such as the Me 262.[34]

In the Pacific theatre, however, the situation proved more fluid. Unlike the RAF, the Admiralty remained staunchly in favour of the Franks suit. During 1944–45, the FAA issued several thousand examples to carrier pilots deploying with the British Commonwealth Pacific Fleet. Its heat and weight problems remained pressing in the tropics, but the need for greater manoeuvrability against Japanese types generally trumped these concerns—at least on operations. The FFS was certainly better suited to the service's roomy Hellcats and Vought F4U Corsairs than to later-mark Seafires. As the USN carrier force grew to staggering proportions, its own Hellcat and Corsair pilots enjoyed the new lightweight, air-operated Z-2 and Z-3 suits introduced in mid-1944. Into 1945, these garments were also adopted as standard by US Marine Corps aviators, both for air-to-air and air-to-ground sorties. Yet, in contrast with naval operators, USAAF commanders in the Pacific remained averse to anti-G gear, emphasising airspeed and height over turning manoeuvres when dealing with Japanese fighters.[35]

Recognising that its Spitfire VIIIs were far more nimble than the Americans' Kittyhawks, Thunderbolts, and Lockheed P-38 Lightnings, the RAAF persevered with plans to deploy CAAG equipment for the island campaigns of late 1944. The Australians also proved far more diligent than the RAF in preparing units for anti-G operations. All pilots in Nos. 452, 548, and 549 Squadrons underwent an anti-blackout indoctrination course on The University of Sydney centrifuge, run by 2 Flying Personnel Research Unit. A detailed manual on the principles, elements, and operation of the Cotton suit also went through several editions, aimed not only at aviators but also fitters, equipment officers, and medical officers.[36] The Cotton suit was also actively promoted by Wing Commander Clive Caldwell, the highly regarded fighter pilot who led the RAAF's No. 80 (fighter) Wing. Finally, another technical exchange saw Monty Cotton posted home to Australia, where he joined 1 Aircraft Performance Unit (APU) at RAAF Base Laverton in October 1944.[37]

Monty Cotton was even better qualified to advise the RAAF in capitalising on his cousin's invention than Carr. Nobody throughout the British Empire could offer a more comprehensive in-theatre experience with anti-G suits in the context of modern Japanese aircraft and tactics. Working with the RAAF's primary anti-G test pilot, Flight Lieutenant Ken Robertson, the 1 APU team soon developed a suite of principles, formations, and manoeuvres for air combat use. Their work was observed by yet another overseas expert. Edward Baldes, an aeromedical research leader from the Mayo Clinic, visited Australia in October–November 1944. He assessed that the RAAFs decision to form a dedicated anti-G test and development flight put Australia ahead of the Allies. In fact, Baldes confirmed, 'Nowhere else has there been any attempt to devise tactics which will combine the advantages of the equipment with the present restriction on tactics brought about by inferior manoeuvrability of allied aircraft'.[38]

Despite this high praise, it was too late for the CAAG programme. Anti-G equipment was systematically removed from all Spitfires under RAAF command before their deployment overseas in December 1944. Perhaps reflecting both a personal dislike for 'gadgets' and pointed feedback from his pilots, Air Vice Marshal Adrian Cole had voiced implacable opposition to the CAAG suit since his appointment as Air Officer Commanding North-Western Area Command in July 1943. 'The disadvantages of the suit still outweigh the advantages afforded by the elevation of a blackout threshold', Cole again insisted in August 1944. In particular, he cited its cumbersome and uncomfortable design, difficulty achieving a good fit, excessive perspiration, and a lack of modification to parachute harnesses which posed a 'definite danger of injury to the testicles in the event of bailing out'.[39]

Thus emasculated, the CAAG suit was soon relegated to reserve status and downgraded from secret to confidential status on 19 January 1945. Its only remaining value to the RAAF was as a propaganda story, ironically promoting the nation's technical prowess at exactly the moment it had been trumped by American pragmatism.[40]

Failings and frailties

With No. 80 Wing's Spitfire units now based on Morotai in the Dutch East Indies, air-to-air combat was rare, but strafing and fatigue remained G-related operational concerns. By this time, 1 APU was well advanced in its evaluation of the US G-3 suit and its associated M-2 valve, both of which had been supplied by Baldes during his Australian tour. Having endured the rubberised Franks and Cotton suits for years, Robertson and Monty Cotton were delighted with this US anti-G outfit, lauding its lightness, flexibility, and breathable nylon construction. They soon prepared tactical recommendations for its use in Spitfires and the RAAF's forthcoming Mustangs, prompting an order for 1245 G-3 suits under Lend-Lease arrangements.[41] However, despite the free flow of anti-G experts, reports, and prototypes, none of the ordered garments reached Australia by the end of the Pacific War.

This was not the final obstacle in the RAAF's attempt to enhance its air power capabilities through technical innovation. In November 1944, Monty Cotton had anticipated future developments in aircraft propulsion and design, such that, 'in an aerial combat the new fighter will only be as manoeuvrable as the tolerance to "G" of the man inside it'.[42] For the present, however, he and Robertson continued their anti-blackout studies in 1 APU's Spitfire VIIIs. In addition to adding warning lights and buzzers to limit overstressing the airframe, the RAAF also decreed in 1943 that Spitfire pilots wearing the anti-G suits should not exceed 6.5 G—even in combat. Out of curiosity, the 1 APU team investigated whether high G affected the aircraft's weapons. 'To our consternation', Monty Cotton later recalled, 'we found that the [.303 inch] machine guns jammed at 5.6 "G" and the [20 mm] cannons at 6.0 "G". So it was back to the drawing board before the suit could be released to the squadrons'.[43] It never was.

Despite twice outfitting its frontline fighter units with a sophisticated suite of anti-G apparatus, the operational benefit for the RAAF was ultimately nil. Only on one occasion—possibly—was it tested in combat with Japanese fighters. Yet the Australians, at least, prepared thoroughly for the service deployment of this novel equipment, proving to be ahead of other Allied nations until the CAAG programme was abandoned. The RAF, in contrast, squandered the early lead offered by the Admiralty's positive combat experience, allowing the production of the Franks suit to escalate while neglecting critical aspects of its implementation.

At the centre of these organisational and operational failings were the frailties of the human body. By focusing too heavily on maximising G tolerance as a physiological and technological problem, the Australian and Canadian inventors underestimated the psychology of flying and the sociology of organisations. If both British and American naval aviators proved more receptive to anti-G suits than air force flyers, the true breakthrough in the US programmes was that medical researchers and manufacturers actually listened to pilots. By providing anti-G suits that aviators would willingly wear, they ensured the mass production and deployment of this technology, albeit at a moment when Allied air supremacy was already assured.

Furthermore, while their suits offered a genuine contribution to air power capabilities, British airframes and armament proved unequal to intensifying the energy at which air combat occurred. Perhaps it is apt to leave the final words with Carr, the wry sojourner whose travels exemplified both the value and the disappointments of wartime technical exchange. 'Until medical research establishes a limiting human factor', he had enthused in 1943, 'it would appear that the pilot may now have greater resistance to the effects of acceleration than the aircraft'.[44]

Notes

1 Report, S. Carr, " 'Cotton' anti-'G' suit," 22 June 1943, National Archives of Australia (NAA), A1196 33/501/24.
2 This garment is unrelated to the 'Sidcot' flying suit invented by Australian Sidney Cotton in 1917.
3 Wilfrid H. Brook, "The Development of the Australian Anti-G suit," *Aviation, Space, and Environmental Medicine* 61, no. 2 (1990): 176. This chapter is dedicated to the late Wilfrid Brook, who generously shared his research materials and insights with the author.
4 S. J. Carr, *You Are Not Sparrows: a Light-hearted Account of Flying Between the Wars* (London: Ian Allan, 1975), 131.
5 Report, Frank Stanley Cotton, "Report on Research of F. S. Cotton, D. Sc. for year 1940 to September 30th," 30 September 1940, University of Sydney Archives (UoSA), P147, Item 3.
6 Peter Hobbins, "The Pigeonhole Waltz: Deflating Innovation in Wartime Australia," *Record* (2015): 3–11; Peter Hobbins, "Spin Doctor: Developing and Deflating the Australian Anti-G suit," *Aviation Heritage* 48, no. 2 (2017): 63–69; Peter Hobbins, "Unearthing Airspace: The Historical Phenomenology of Aviation Artefacts," *Australasian Historical Archaeology*, no. 37 (2019): 43–55; Peter Hobbins, "Engineering the Fighter Pilot: Aviators, Anti-G suits, and Allied Air Power, 1940–53," *Journal of Military History* 84 (2020): 115–49.
7 Bulletin, "Tactical bulletin No. 18/43," 31 August 1943, NAA, A705, 132/1/830.
8 Report, J.H. Bisley, "First impressions of 'G' suit,' n.d. [c. July 1943], NAA, A705, 43/1/586.
9 Hobbins, "Engineering the Fighter Pilot," 130.
10 Norman L.R. Franks, *The Air Battle of Imphal* (London: William Kimber, 1985), 21–35; Norman Franks, *Aircraft Versus Aircraft: The Illustrated Story of Fighter Pilot Combat Since 1914* (New York: Macmillan Publishing Company, 1986), 108–45.

11 Peter Allen, "The Remotest of Mistresses: the Story of Canada's Unsung Tactical Weapon: the Franks Flying Suit," *CAHS Journal* 21 (1983): 110–21, 126; George Smith, "The Franks Flying Suit in Canadian Aviation Medicine History, 1939–1945," *Canadian Military History* 8, no. 2 (1999): 35–41.

12 Report, J. Dunn, "Policy – Franks Flying Suit," 22 November 1943, The National Archives (UK) (TNA), AIR 20/2134.

13 Report, C. H. Kellaway, "Report of Medical Scientific Mission to USA Canada and United Kingdom, 29 September 1941 to 16 April 1942," 24 April 1942, Australian War Memorial (AWM), AWM54, 481/12/164; Peter Hobbins and Hannah Forsyth, "Mobilising Medical Knowledge for the Nation, 1943–49," *Health and History* 15, no. 1 (2013): 66–67.

14 Letter, Charles H. Kellaway to Dr Cotton, 11 September 1941, UoSA, P147, Item 26.

15 Liam Kane, "Allied Air Intelligence in the South West Pacific Area, 1942–1945," *Journal of Intelligence History* (2021): 1–21; Frank Cain, "The Role of Scientific Intelligence in the Pacific War," in *Science and the Pacific War: Science and Survival in the Pacific, 1939–1945*, ed. Roy MacLeod (Dordrecht: Kluwer Academic Publishers, 2000), 271–89; Walter V. Abraham, *AIRIND in Retrospect: a Royal Australian Air Force Contribution to Allied Intelligence, Pacific War 1942–1945* (Kiama: Saddleback Press, 1996).

16 Allan S. Walker, *Australia in the War of 1939–1945: Series Five, Vol. IV: Medical Services of the R.A.N. and R.A.A.F.* (Canberra: Australian War Memorial, 1961), 351–52; Maurice W. Kirby, "Operations Research Trajectories: the Anglo-American Experience from the 1940s to the 1990s," *Operations Research* 48, no. 5 (2000): 661–70.

17 Report, S/Ldr E.A. Pask, n.t., n.d. [c. July 1945], NAA, A705, 132/12/20.

18 Report, "Outline of duties of personnel at Nos. 1 and 2 F.P.R.U.," n.d., NAA, A705, 231/9/1634.

19 Walker, *Medical Services of the R.A.N. and R.A.A.F.*, 352–57.

20 Database, Flying Personnel Research Committee – unclassified reports, Farnborough Air Sciences Trust, 2014. I am grateful to Graham Rood for sharing this database with the author.

21 Report, S. J. Carr, "Report on comparative trials between C.A.A.G. and F.F. suits," 11 February 1944, NAA, A705, 132/1/829.

22 Peter Hobbins, "Flying the 'Bathfire': Britain's Wartime 'Anti-g Suit' Research and Development," *The Aviation Historian,* no. 32 (2020): 92, 96–98.

23 Report, M. C. C. Cotton, "F.F.S.," 6 December 1943, Library and Archives Canada (LAC), RG24, 5361, 45–2–18.

24 Cotton, 'F.F.S.'

25 Report, J. Dunn, "Frank's Flying Suit, Mark II," 30 January 1944, TNA, AIR 20/2134.

26 Report, D. of Ops (A.D.), "Frank's Suit – policy," 26 January 1944, TNA, AIR 20/5193.

27 David Stubbs, "A Blind Spot? The Royal Air Force (RAF) and Long-range Fighters, 1936–1944," *Journal of Military History* 78, no. 2 (2014): 689–90.

28 For early footage demonstrating the FFS, see "A Film About the Effects of G-Force,' c.1941, Huntley Film Archive, film 1004471, www.huntleyarchives.com/preview.asp?image=1004471&itemw=4&itemf=0001&itemstep=1&itemx=1.

29 M.C. "Bush' Cotton," *Hurricanes Over Burma: the Story of an Australian Fighter Pilot in the Royal Air Force* (Oberon: Titania Publishing, 1988), 301–10.

30 Minutes, "Minutes of thirteenth meeting of the Flying Personnel Research Committee," 31 August 1944, NAA, MP1049/5, 1968/2/564.

31 Report, Kenneth E. Penrod, "Installation of anti-G equipment in the Eighth Air Force fighter groups," 10 April 1945, National Air and Space Museum Archives, M6–301320–01.

32 Kerrie Dougherty, "Australia's Contributions to the Early Development of the Partial-pressure Suit: A History of the Cotton Anti-G Suit," in *Proceedings of the Eleventh*

National Space Engineering Symposium, ed. Insitution of Engineers, Australia (Sydney: Institution of Engineers, 1997), 89–98.
33 David R. Jones, *Flight Surgeon Support to United States Air Force Fliers in Combat* (Brooks City: USAF School of Aerospace Medicine, 2003), 25–41.
34 Dennis R. Jenkins, *Dressing for Altitude: U.S. Aviation Pressure Suits – Wiley Post to Space Shuttle* (Washington, DC: National Aeronautics and Space Administration, 2012), 89–109.
35 Hobbins, "Engineering the Fighter Pilot," 140–43.
36 Manual, "C.A.A.G. Manual," n.d. [c.1944], Royal Australian Air Force Museum (RAAFM), ME/43.
37 Presentation, Monty Cotton, "The Development of the Anti-Blackout Suit in World War Two," 28 February 1997, RAAFM, Monty Cotton Lecture 1997.
38 Report, E.H. Anderson, "Report on visit to New Guinea in company with Dr. E.J. Baldes," 21 January 1945, NAA, A705, 39/5/1362.
39 Letter, A.T. Cole, "C.A.A.G. equipment – Mark I (modified)," n.d. [c. 14 August 1944], NAA, A705, 132/1/829.
40 "Spitfire Pilot's 500 miles Per Hour Dive in Anti-blackout Suit," *Sun* (Melbourne), 28 April 1945, 12–13.
41 Minute, George C. Ellis, "Minute 24," 13 July 1945, NAA, A705, 39/5/1362.
42 Report, M. C. C. Cotton, "Comments on anti-G work, No. 2 F.P.R.U.," n.d. [c. November 1944], NAA, A1196, 33/501/36.
43 Cotton, "The development of the anti-blackout suit."
44 Report, S. Carr, " 'Cotton' anti-'G' suit," 22 June 1943, NAA, A1196, 33/501/24.

References

Archival sources

Australian war memorial

AWM54: Written records 1939–45 War.

Library and Archives Canada

RG24: Department of National Defence fonds.

National Air and Space Museum Archives – Smithsonian (Washington D.C.)

M6–301320–01: General, WWII, G-suits.

Royal Australian Air Force Museum

ME/43: Medical Services.
Monty Cotton Lecture, 1997.

The National Archives of Australia

A1196: Department of Air, Central Office, Correspondence files, 1935–1960.

A705: Department of Defence, Correspondence files, 1912–1988.
MP1049/5: Navy Office, Correspondence files, 1923–1950.

The National Archives (UK)

AIR 20: Air Ministry and Ministry of Defence: Papers accumulated by the Air Historical Branch.

University of Sydney Archives (Sydney)

P147: Personal Papers of Frank Stanley Cotton.

Published sources

Abraham, Walter V. *AIRIND in Retrospect: A Royal Australian Air Force Contribution to Allied Intelligence, Pacific War 1942–1945.* Kiama: Saddleback Press, 1996.

Allen, Peter. "The Remotest of Mistresses: The Story of Canada's Unsung Tactical Weapon: The Franks Flying Suit." *CAHS Journal* 21 (1983): 110–21, 126.

Brook, Wilfrid H. "The Development of the Australian Anti-G Suit." *Aviation, Space, and Environmental Medicine* 61, no. 2 (1990): 176–82.

Cain, Frank. "The Role of Scientific Intelligence in the Pacific War." In *Science and the Pacific War: Science and Survival in the Pacific, 1939–1945,* edited by Roy MacLeod, 271–89. Dordrecht: Kluwer Academic Publishers, 2000.

Carr, S. J. *You Are Not Sparrows: A Light-Hearted Account of Flying between the Wars.* London: Ian Allan, 1975.

Cotton, M. C. "Bush." *Hurricanes Over Burma: The Story of an Australian Fighter Pilot in the Royal Air Force.* Oberon: Titania Publishing, 1988.

Dougherty, Kerrie. "Australia's Contributions to the Early Development of the Partial-Pressure Suit: A History of the Cotton Anti-G Suit." In *Proceedings of the Eleventh National Space Engineering Symposium,* edited by the Institution of Engineers, Australia, 89–98. Sydney: Institution of Engineers, 1997.

Franks, Norman L. R. *The Air Battle of Imphal.* London: William Kimber, 1985.

———. *Aircraft Versus Aircraft: The Illustrated Story of Fighter Pilot Combat Since 1914.* New York: Macmillan Publishing Company, 1986.

Hobbins, Peter. "The Pigeonhole Waltz: Deflating Innovation in Wartime Australia." *Record* (2015): 3–11.

———. "Spin Doctor: Developing and Deflating the Australian Anti-G Suit." *Aviation Heritage* 48, no. 2 (2017): 63–69.

———. "Unearthing Airspace: The Historical Phenomenology of Aviation Artefacts." *Australasian Historical Archaeology,* no. 37 (2019): 43–55.

———. "Engineering the Fighter Pilot: Aviators, Anti-G Suits, and Allied Air Power, 1940–53." *Journal of Military History* 84 (2020a): 115–49.

———. "Flying the 'Bathfire': Britain's Wartime 'Anti-G Suit' Research and Development." *The Aviation Historian,* no. 32 (2020b): 92–103.

Hobbins, Peter, and Hannah Forsyth. "Mobilising Medical Knowledge for the Nation, 1943–49." *Health and History* 15, no. 1 (2013): 59–79.

Jenkins, Dennis R. *Dressing for Altitude: U.S. Aviation Pressure Suits – Wiley Post to Space Shuttle.* Washington, DC: National Aeronautics and Space Administration, 2012.

Jones, David R. *Flight Surgeon Support to United States Air Force Fliers in Combat.* Brooks City: USAF School of Aerospace Medicine, 2003.

Kane, Liam. "Allied Air Intelligence in the South West Pacific Area, 1942–1945." *Journal of Intelligence History* (2021): 1–21.

Kirby, Maurice W. "Operations Research Trajectories: The Anglo-American Experience from the 1940s to the 1990s." *Operations Research* 48, no. 5 (2000): 661–70.

Smith, George. "The Franks Flying Suit in Canadian Aviation Medicine History, 1939–1945." *Canadian Military History* 8, no. 2 (1999): 35–41.

Stubbs, David. "A Blind Spot? The Royal Air Force (RAF) and Long-Range Fighters, 1936–1944." *Journal of Military History* 78, no. 2 (2014): 673–702.

Walker, Allan S. *Australia in the War of 1939–1945: Series Five, Vol. IV: Medical Services of the R.A.N. and R.A.A.F.* Canberra: Australian War Memorial, 1961.

Part Four

Air Power in the 21st Century

13 Ethics, strategy, and Australian air power in the 21st century

Deane-Peter Baker

In his influential *Military Review* paper, Colonel Arthur F Lykke, Jr famously divided the elements of strategy into 'ends, ways, and means'.[1] Potentially, Air power provides, one means available to the strategist. Of course, these three elements of strategy are interconnected. In an ideal world, the ends that are sought will define what ways are employed to achieve those ends, and the ways will, in turn, dictate what means are required. In practice, however, there is often strong pressure in the other direction—the means available give rise to a limited set of possible ways, which in turn can constrain the range of ends that are realistically available. It takes little imagination to realise that there is considerable danger in strategy driven by capabilities that have not been developed with due consideration of their relationship to appropriate ends. Air power is one of the most significant capabilities available to those formulating and executing strategy on behalf of the Australian nation, and it is, therefore, both encouraging and important that this book evaluates the strategic role of Australian air power in the 21st century. In this chapter, I will consider the role of ethics in that evaluation.

From the outset, it is important to acknowledge that (to understate the matter somewhat) not every strategist thinks that ethics should have a role in strategy formulation; indeed, some are vehemently opposed to the idea. In a famous passage in the most famous book on strategy, *On War*, Carl von Clausewitz argues forcefully against incorporating ethical considerations, particularly 'benevolence born of philanthropy', into strategic thought. He contends:

> [I]n such dangerous things as War, the errors which proceed from a spirit of benevolence are the worst. As the use of physical power to the utmost extent by no means excludes the co-operation of the intelligence, it follows that he who uses force unsparingly, without reference to the bloodshed involved, must obtain a superiority if his adversary uses less vigour in its application.[2]

Admittedly, Clausewitz's position on ethics is—like most topics addressed in *On War*—complex and somewhat difficult to pin down, and this quote alone does not necessarily reflect the full picture. It does, however, give us a sense of why many strategists are leery of discussions of ethics. 'War', we might imagine them saying, 'is too important to be left in the soft pale hands of ethicists'. In this chapter,

DOI: 10.4324/9781003230656-18

I hope to go some way to changing that view by showing how ethical considerations ought to be intrinsic to thinking about the strategic use of air power in the 21st century. I begin with a discussion of the problems caused by what Conrad Crane calls 'the lure of strike' to illustrate the dangers of means-driven strategy. I then turn to the broader issue of the role ethics plays in strategy and suggest a revision that provides—I think—a basis on which to reimagine air power for the 21st century.

The lure of strike

There is no doubt that the technology-driven capability that states such as the United States (US) (and, to a lesser degree, smaller developed nations such as Australia) have to project force through air power represents a massive asymmetrical advantage over the non-state actors and developing world states which have been the primary opponents in recent military clashes. Dr Conrad Crane—historian, strategist and air power specialist—is a great fan of asymmetrical advantage. As he is fond of saying, 'there are two ways to wage war, asymmetric and stupid'.[3] Nonetheless, Crane (who was entrusted by his former West Point classmate David Petraeus with the task of leading the team that developed the ground-breaking US Army and Marine Corps counterinsurgency manual, which was released in December 2006) has expressed strong concern over the employment of US and allied air power in recent decades. At the heart of that concern is what Crane calls 'the lure of strike'. As he explains:

> An increasingly important part of the New American Way of War has been a reliance on standoff technology to project power. The 'lure' is minimal friendly casualties and short, inexpensive wars with only limited land power commitments. Unfortunately, inflated expectations for such an outcome have often led to strategic overreach and a dangerously unbalanced force structure.[4]

If we reassemble Crane's critique within Lykke's ends, ways, and means framework, a problem emerges. The unrivalled ability to project force that air power provides to the US offers such an enticing means to strategists that it is distorting the process of strategic decision making to the point that ways are being undermined and ends not achieved. It is noteworthy in the aforementioned quote that Crane takes the ability to project force as being synonymous with the ability to project power—I will return to this issue below. Worse still, in Crane's view, this distortion has the further effect that the nation's force structure—the sum of US military means—is being pushed further toward a reliance on offensive air power, and a vicious circle is in effect.

I wish to add another aspect to this analysis. Overlooked in the quest for 'short, tidy wars' is the fact that these wars are rarely, if ever, particularly tidy for those on the ground.[5] For all the—enormously impressive—advances that we have seen in the development of precision air-delivered munitions, the fact remains that air strikes are not instances of what my former colleague at the US Naval Academy,

Stephen D Wrage, calls (with no small irony) 'immaculate warfare'.[6] We simply cannot pretend that precision munitions take collateral damage out of the equation, especially in an era of global telecommunication in which collateral damage is beamed into living rooms and to handheld devices across the globe on an almost daily basis. Of course, the fact that Australia has generally been a coalition partner to the US in recent conflicts means that the Australian Government has the luxury of being more selective about the targets selected to be 'serviced' by the Royal Australian Air Force. But while this gives a degree of political cleanness to these operations, the fact is that in these conflicts, we are willing partners in a campaign that is killing innocent people on the ground.

By pointing this out, I do not intend to suggest that we should only ever employ armed force where we can be sure no collateral damage will occur—that would be tantamount to a declaration of contingent pacifism.[7] The point here is that one aspect of the lure of strike is the attractiveness of the misperception that air power provides a means to wage immaculate warfare. It does not—and we should not kid ourselves or mislead the public into thinking that it does.

The issue of collateral damage puts the doctrine of double effect (DDE) squarely on the table. The DDE is the mechanism in ethics used to discern legitimate from illegitimate acts that involve negative consequences for those who are, in the relevant sense, innocent. The DDE provides for the act in question (such as a proposed air strike) the following four-step test:

1. The act itself (i.e., the act that will result in the harm in question) is either a good or morally neutral act.
2. The act is intended to achieve the good effect that will result, not the bad effect. The test here is a counterfactual—would the act go ahead if the bad effect were not going to occur?
3. The good effect must not be caused by the bad effect.
4. The harm caused by the act must not be out of proportion to the good that will be achieved.[8]

The DDE enables decision-makers to give legitimacy to strikes that will cause 'foreseen, but unintended' deaths among non-combatants. It is also among the most challenged ethical doctrines. The most powerful objection to the doctrine is that it places too much emphasis on intentions. While it is a good thing that non-combatants are not intentionally targeted, it matters little to those who are killed, or to their loved ones, that their deaths were foreseen but not intended, rather than directly intended. So, is this a strong enough basis on which to rest a distinction between the permissible and impermissible killing of non-combatants? Air power, arguably above all other forms of armed force available to contemporary strategists, places the greatest weight on this fragile doctrine.

But this is not the biggest ethical challenge that arises because of the lure of strike. As already discussed, the tendency is for the lure of strike to apply pressure upstream, with the consequence that ends are shaped (or, perhaps better, misshaped) by means. Air power, popularly understood (as it usually is in the contemporary environment at the expense of critical air power contributions such as air mobility and intelligence, surveillance, and reconnaissance), is essentially about the delivery of explosive warheads from aerial platforms. Those who subscribe to this view see air power as a very effective way of applying violence to opposing forces, however not much good at doing anything else. Consequently, there is a strong temptation to see the end as being about (as our US allies would say) 'attriting' the enemy. We become thereby focused on a game of 'whack-a-mole', striking at the enemy wherever he becomes visible. As we view the conflict through the soda straw of uncrewed aerial vehicle (UAV)-delivered imagery, we lose sight of our ethical responsibilities, those that go beyond merely doing what we can to make sure we're targeting the right people thousands of feet below. It becomes too easy to overlook our broader responsibilities to those who live on the ground we are flying over and dropping bombs on. That is because fulfilling those responsibilities will often require putting boots on the ground. There is some acknowledgement of this in the current approach in some conflicts to make use of (to use a current buzz phrase) 'indigenous mass' to achieve the broader objectives that our Western liberal values demand when we dare to glance up at them. But this 'solution' generates a raft of further additional problems. Can we trust our indigenous surrogates to do what should be done? Rarely do they share our values or the training needed to ensure those values are acted on in the heat of battle. They often carry histories, agendas, and prejudices that can spill over into abuses and atrocities carried out against the very innocents we are seeking to protect. What responsibility do we then bear? It will not do to simply shrug our shoulders and turn back to the UAV feed.

Rethinking ethics and strategy

The late US Air Force General Curtis Lemay, infamous for designing the strategic bombing campaign in the Pacific theatre during the Second World War, is famously quoted as having said, 'If we maintain our faith in God, love of freedom, and superior global air power, the future looks good'.[9] It is a sentiment that falls awkwardly on contemporary ears. The idea of appealing to God as a justification for engaging in war is something 'they' do, not something 'we' do.[10] But as awkward as it sounds, Lemay's statement is a clear example of what has for a very long time been the dominant way of thinking about the ethics of armed conflict, what we might call 'top-down justification'. Here the idea is that it is *la grande cause* that justifies the acts of war, from the strategic level down to the tactical. In its extreme form—as is arguably the case for Lemay's justification for strategic bombing—*la grande cause* is taken to be weighty enough to justify a very wide range of actions and gets very close to the amoralism of hard-nosed realism. In a more moderate form, *la grande cause* (duly moderated by the constraints of the *jus ad bellum*) justifies the ends sought, and the ways and means are left to considerations of military effectiveness, so long as those ways and means do not overstep the thin red lines of the *jus in bello* principles of necessity, proportionality, and discrimination/distinction.

However, in recent years, what was once the dominant picture has changed. In part, a result of the intensification of the value placed on individuals in liberal societies, and in part, because of the growing distaste for grand narratives in the postmodern world, the thin red line of ethical constraint that previously constrained military action has grown thicker and thicker, increasingly narrowing the scope of military action, to the point where many military professionals chafe at the 'noose' of ethical (and, increasingly, legal) limitations. Consider, for example, the following passage from the book *No Easy Day*, an account of the career of a US Navy SEAL, Matt Bissonnette, writing under the pseudonym Mark Owen, who took part in the raid to kill Osama bin Laden. Before that mission, Operation Neptune Spear, Bissonnette had been deployed to Afghanistan on multiple tours. But he found over time that things seemed to be changing:

> For years we had been sneaking into compounds, catching fighters by surprise. Not anymore.
> On the last deployment, we were slapped with a new requirement to call them out. After surrounding a building, an interpreter had to get on a bullhorn and yell for the fighters to come out with their hands raised. It was similar to what the police did in the United States. After the fighters came out, we cleared the house. If we found guns, we arrested the fighters, only to see them go free a few months later. Often, we recaptured the same guy multiple times during a single deployment.
> It felt like we were fighting the war with one hand and filling out paperwork with the other. When we brought back detainees, there was an additional two or three hours of paperwork. The first question to the detainee at the base was always, 'Were you abused?' An affirmative answer meant an investigation and more paperwork.
> And the enemy had figured out the rules.[11]

In a similar vein, in his book *The Crossroad*, Victoria Cross for Australia recipient and Special Air Service Regiment operator Corporal Mark Donaldson expresses his bewilderment over an interaction with a senior officer after a firefight in which he and his combat assault dog, Devil, were involved:

> Some time after the big fight, I was training with Devil. A senior officer was asking me why, on that day, so many people had had to die. I had a simple answer.
> 'Sir, they were trying to kill us'. I explained in detail about being in that house, at close quarters, with those guns firing at us out of rooms. I explained what Devil had done, and how I'd shot the man who was fighting with Devil while trying to get his gun aimed at me.
> The officer said, 'Did you try to detain him?'
> I was dumbounded. The question showed a complete lack of understanding. We were in a war situation, not a policing situation. There's often a lot of exaggerated talk about 'Kill or be killed', but this time, that was what it was.

If I, or others that day, had not killed, then we wouldn't be here today, and my children would not have a dad.[12]

Reflected in both accounts is a general sense of frustration that, it seems to me, has seeped through the ranks of modern Western forces. It is a frustration that emerges from the mismatch between the approach to the use of military force that arose under the top-down justification paradigm, on the one hand, and the growing ethical and legal strictures that contemporary sensibilities are placing on military action.

So, what is the solution? I believe that it simply will not do, as many military practitioners would like, to 'push back' against the encroachment of greater ethical demands and military legalism to defend the traditional domain of operational art. That will never resolve the issue, and it is only likely to lead to greater frustration and a growing rift between the military and the society it serves. Instead, I believe it is necessary to radically reimagine the relationship between ethics and strategy. In an important recent book, the eminent military ethicist George Lucas argues that the nature of 21st-century armed conflict, particularly the rise of irregular opponents, has moved ethics from the cold periphery to the very centre of sound military strategy.[13] It is a view I agree with wholeheartedly. As far back as 2007, I co-authored, with Dr Deborah Roberts, the first of a brace of papers exploring the possibility of repurposing one of the most influential ethical frameworks applied to the realm of development policy—Martha Nussbaum's capabilities approach—to be the key driver in developing military strategy.[14] Whether adapting the capabilities approach is the best approach to reimagining the relationship between ethics and strategy is a question that is beyond the scope of this particular chapter, but the general point, it seems to me, is correct: the ethics of armed conflict must move from being a set of perimeter constraints to military action to become the core driver instead. That is, it seems to me that the best response to the inadequacies of top-down justification is not to try to counter it with 'bottom-up' ethics. The 'outside-in' application of ethics to military decision making that is currently the norm must be replaced with an 'inside-out' approach in which ethical considerations are the starting point rather than a compliance check at the end.

Rethinking air power for the 21st century

I started this chapter by highlighting the dangers of having the standoff targeting capability of contemporary air power drive strategy by shaping ends. How might we rethink air power in the light of the proposed reimagining of the relationship between ethics and strategy described earlier? I do not have a complete answer to that question—indeed, developing a satisfactory answer will take far more than the formulation of ideas by any single military ethicist. However, it does seem to me that a critical starting point will have to be to challenge a fundamental assumption at the heart of the military profession: the idea that the core business of the military is the application of force, and everything else is secondary.

Lucas points out that Clausewitz—whose shadow touches all contemporary strategy—is a man of his time and deeply, even inescapably, influenced by the dominant paradigm of his day, classical dynamics, also known as Newtonian mechanics. To Clausewitz, militaries are, in essence, complex and highly tuned industrial era machines that target centres of 'gravity' to overcome the opposing 'vectors' of 'inertia' using 'force' to achieve political ends.[15] In Clausewitz's Newtonian paradigm, force is the cornerstone on which everything else rests. That this is true also of how contemporary militaries conceive of their role came home to me forcefully a few years ago when I attended a briefing in which an officer went through several PowerPoint slides outlining the 'non-kinetic fires' that had been available to him in theatre. Indeed, so powerful is this core assumption that it is difficult to think of alternative descriptors to replace the idea of a military 'force'.

It is, I admit, hard to imagine what (to coin a phrase) a 'non-Newtonian military' might look like. Even narrowing the scope to focus only on how non-Newtonian air power might look gives us no easy answers. What seems necessary is a broadening of scope in the minds of those who default to the application of kinetic effect as the only valuable air power effect. One only needs to review the backgrounds of the RAAF's senior leadership team, disproportionately (compared with their non-kinetic colleagues) populated as it is by fighter pilots, to observe the relative value of non-kinetic air power. A non-Newtonian air force could look past the lure of strike to see how diverse air power contributions might achieve national strategic effect. For instance, UAVs capable of delivering medical care to civilians trapped in conflict zones, aerial platforms with the capability to observe, record, and transmit, in real time, the activities of fighters on the ground air-launched less-lethal weapons, aerial fire-fighting and even cloud-seeding capabilities have, depending on the circumstances, the capacity to deliver meaningful strategic outcomes that far outweigh the destructive power of kinetic munitions; however, they are generally viewed as secondary 'supporting effects' by those trapped in the contemporary Newtonian paradigm. None of this is to suggest that the ability to deliver ordnance from the air has somehow suddenly become superfluous—of course, that is not even *close* to being true—but it seems to me that this fact does not negate the need for us to open ourselves to rethinking just what we value as air power.

Developing an ethics-driven, non-Newtonian conception of air power will take considerable imagination. It will also require us to challenge a good number of deeply rooted assumptions, many of which we have only begun to recognise. Undoubtedly that will be a process that will be unsettling and will meet with considerable resistance, but if we are unwilling to face up to that challenge, we will find the scope of military action to be increasingly deadlocked by the opposing forces of top-down and bottom-up justification.

Notes

1 Arthur F. Lykke, "Defining Military Strategy," *Military Review* 69, no. 5 (May 1989): 2–8.

2 Carl von Clausewitz, *On War: Book I* (London: Routledge, 1827), Chaps 1–2.
3 Conrad C. Crane, "Special Commentary: The Lure of Strike," *Parameters* 43, no. 2 (Summer 2013), 5.
4 Ibid., 5.
5 Ibid., 5.
6 See: Stephen D. Wrage, ed., *Immaculate Warfare: Participants Reflect on the Air Campaigns Over Kosovo, Afghanistan, and Iraq* (Westport, CT: Praeger, 2003).
7 A contingent pacifist is, roughly speaking, someone who agrees in principle with Just War Theory, but believes that no war can, or does, in practice meet the requirements for a just war, hence no war is legitimate. For the most comprehensive analysis of contingent pacifism to date, see: Andrew Alexandra and Ned Dobos, *The New Pacifism: Just War in the Real World* (Oxford: Oxford University Press, 2018).
8 Deane-Peter Baker, "Collateral Damage and the Doctrine of Double Effect," in *Key Concepts in Military Ethics*, ed. Deane-Peter Baker (Sydney: NewSouth Publishing, 2016), Chap 25. EBSCOhost eBook.
9 Quoted from James Streckfuss, *Eyes in the Sky: Aerial Reconnaissance in the First World War* (Oxford: Casemate, 2016), 156.
10 Christopher Eberle offers a rare contemporary treatment of the ethics of armed conflict from a Western perspective which defends the legitimacy of religious reasons in judgements of war. See: Christopher J. Eberle, *Justice and the Just War Tradition: Human Worth, Moral Formation and Armed Conflict* (London: Routledge, 2016).
11 Mark Owen and Kevin Maurer, *No Easy Day: The Navy SEAL Mission that Killed Osama Bin Laden* (London: Penguin, 2012), 155–56.
12 Mark Donaldson, *The Crossroad: A Story of Life, Death and the SAS* (Sydney: Pan Macmillan, 2013), 375.
13 George Lucas, *Ethics and Military Strategy in the 21st Century: Moving Beyond Clausewitz* (Oxon: Routledge, 2019).
14 Deane-Peter Baker and Deborah Roberts, "Extending Just War Theory: The Jus Ad Bellum and the Capabilities Approach to Armed Conflict," *Scientia Militaria* 35, no. 2 (2007): 21–38; Deane-Peter Baker and Deborah Roberts, "Assessing the Capabilities Approach to the Ethics of Armed Conflict: Jus in Bello and the Case of Operation Cobra II," *Politeia* 27, no. 2 (2008): 89–103.
15 Lucas, *Ethics and Military Strategy in the 21st Century*.

References

Alexandra, Andrew, and Ned Dobos. *The New Pacifism: Just War in the Real World.* Oxford: Oxford University Press, 2018.
Baker, Deane-Peter. "Collateral Damage and the Doctrine of Double Effect." In *Key Concepts in Military Ethics*, edited by Deane-Peter Baker, chap. 25. Sydney: NewSouth Publishing, 2016. EBSCOhost eBook.
Baker, Deane-Peter, and Deborah Roberts. "Extending Just War Theory: The Jus Ad Bellum and the Capabilities Approach to Armed Conflict." *Scientia Militaria* 35, no. 2 (2007): 21–38.
———. "Assessing the Capabilities approach to the Ethics of Armed Conflict: Jus in Bello and the Case of Operation Cobra II." *Politeia* 27, no. 2 (2008): 89–103.
Crane, Conrad C. "Special Commentary: The Lure of Strike." *Parameters* 43, no. 2 (Summer 2013): 5–12.
Donaldson, Mark. *The Crossroad: A Story of Life, Death and the SAS.* Sydney: Pan Macmillan, 2013.
Eberle, Christopher J. *Justice and the Just War Tradition: Human Worth, Moral Formation and Armed Conflict.* London: Routledge, 2016.

Lucas, George. *Ethics and Military Strategy in the 21st Century: Moving Beyond Clause-witz*. Oxon: Routledge, 2019.

Lykke, Arthur F. "Defining Military Strategy." *Military Review* 69, no. 5 (May 1989): 2–8.

Owen, Mark, and Kevin Maurer. *No Easy Day: The Navy SEAL Mission that Killed Osama Bin Laden*. London: Penguin, 2012.

Streckfuss, James. *Eyes in the Sky: Aerial Reconnaissance in the First World War*. Oxford: Casemate, 2016.

von Clausewitz, Carl. *On War: Book I*. London: Routledge, 1827.

Wrage, Stephen D., ed. *Immaculate Warfare: Participants Reflect on the Air Campaigns over Kosovo, Afghanistan, and Iraq*. Westport, CT: Praeger, 2003.

14 Manoeuvre in the 21st century

Jo Brick

In an article published by the United States (US) Army's Modern War Institute, Zachery Tyson Brown argued that 'war is a social construct, an interaction between political communities, [and] its expression changes in line with the tools we use to make those interactions'. Brown discusses the forces of societal change—particularly the importance of information as a commodity in society—that underscore what defence forces call 'grey zone' or 'hybrid wars'.[1] This is a challenge because war and peace are generally considered binary, and it is tempting to use military means to deal with most security challenges. Russia is generally seen as an exemplar of state actors who seem to deftly use all methods and means at their disposal, during both peace and war, to exert influence in pursuit of its strategic objectives. Carl von Clausewitz, the Prussian general and military theorist whose work remains central in military theory, wrote that war is a continuation of state policy by other means; however, the Western tendency is to focus on 'war' while other nations focus on the 'continuation of policy by other means', and the Information Age has enabled the pursuit of the latter.[2]

This chapter proposes a different conception of manoeuvre, one that locates it as a unifying concept that brings together all elements of national power to place our nation—as the referent object of security—in a position of relative advantage within the context of perpetual grey zone conflict and competition. If strategy provides us with the goals, manoeuvre—primarily strategic manoeuvre—will be the method to achieve those goals. Having been introduced to the concept of manoeuvre in the context of manoeuvre warfare and air-land battle as one of the central doctrines for joint warfighting in the late 20th century, further consideration of manoeuvre concepts regarding war has led me to believe that it can offer a framework for addressing security challenges beyond warfare. Consequently, this chapter does not discuss manoeuvre *warfare* but rather a different framework for considering manoeuvre concepts to address the challenges faced in the 21st century. If 'grand strategy' provides us with the overarching goal, then perhaps a new conception of manoeuvre can provide a method for synchronising action throughout the necessary government departments to create a truly 'whole of government' approach.

There are four ways a strategic conception of manoeuvre can address contemporary, multifaceted, and persistent security challenges that will be discussed in

DOI: 10.4324/9781003230656-19

this chapter in detail. Firstly, the differences between the Information Age and the Industrial Age are outlined to demonstrate why the use of force alone as a means of solving security problems is ineffective for any long-term solution. This is followed by a brief examination of the origins of manoeuvre theory, which remain relevant today, particularly when considering a coordinated and strategic approach to meeting the challenges of hybrid or grey zone conflict. We can conceive of strategic manoeuvre as a unifying foundation for addressing these threats in times of both peace and war.

This chapter will then discuss why it is necessary to reframe the use of military force in addressing some of these grey zone challenges. It is generally understood that the use of force has limited utility in trying to subdue ideology in the Information Age. Yet, Western nations are still susceptible to what Cathal Nolan calls 'the allure of battle' to address complex security threats.[3] The temptation exists to use military capabilities as a rapid solution to many security problems.

Industrial Age versus Information Age

Changes in technology directly influence warfare in society. The first two decades of the 21st century have witnessed the diffusion of power in the Information Age, which demands a broader approach to addressing security challenges. The Industrial Age was characterised by the mass production of consumer items, military materiel and weapon systems. Indeed, the world wars of the 20th century were won in the factories and capital markets of the world by nation states that could effectively harness industry towards their war effort.[4] However, the 'information revolution' has since eroded industrial might, making it less instrumental in dealing with security challenges. Alvin and Heidi Toffler envisaged this in their work *War and Anti-War*, in which they argue:

> While land, labour, raw materials, and capital were the main 'factors of production' in the Second Wave [or industrial] economy of the past, knowledge—broadly defined here to include data, information, images, symbols, culture, ideology and values—is the central resource of the Third Wave [or information] economy.[5]

The outcome is the 'de-massification' of mass production. Today, we are at a point of having what the Tofflers call a 'split-level economy'—an economy based on declining Second Wave mass production and emergent Third Wave technologies across society generally.[6] This is perhaps reflected in the mix of capabilities within the Australian Defence Force (ADF). Even the Royal Australian Air Force is transitioning towards these Third Wave Technology-based weapon systems, which rely on information for their lifeblood.

The transition between industrial and information-based social systems has eroded the monopoly on industry and capital long-held by states. Coercive effects are no longer reliant on the use of military force. Information has become the new weapon system, as it is a vector for shaping and influencing perceptions of reality and decision-making in a manner that is favourable to one side or another. Information is analysed and processed to create intelligence, which informs commanders' actions. Consequently, opposing forces have sought to shape, deny, manipulate, and alter information to influence decision-making. For instance, one of the better-known deception campaigns of the Second World War, Operation Fortitude, involved Allied deception in the lead up to the invasion of France in 1944. The operation sought to influence German attitudes concerning where an attack would likely occur and, in turn, where the German military would take up defensive positions and position units.[7]

So, what is different today? The free flow and accessibility of information during the Information Age effectively means that an adversary can shape, deny, manipulate, and alter information to influence whole societies. Clausewitz suggested that '[w]ar is thus an act of force to compel our enemy to do our will'.[8] However, states or non-state actors no longer need to rely on force to compel the enemy, as in the Information Age, they can now shape action through the vector of information. Social media is one vector non-state actors have used to shape and influence the information related to them. For instance, P W Singer has explored the use of social media by Daesh, whose invasion of Iraq in 2014 was, Singer argues, launched by a hashtag.[9] Known more commonly as Islamic State, Daesh used social media effectively to influence, convince, and coerce a global audience to support its cause. The group used the information environment to support and complement its operations in the physical space, tweeting about its successes and using all forms of media to push its worldview and rally people to its cause. Daesh can be, and has been, dealt with via military means, but it remains to be seen whether this is a long-term success. The difficulty in engaging with these entities using purely military means is that any success is likely short-term and localised. Long-term solutions that address radicalisation and curb the factors that incentivise people to rally to these organisations do not reside in the military. Instead, enduring solutions to such security challenges require a synchronised approach offered by the new conception of manoeuvre.

US Army Major Marc Romanych provides another perspective on how information shapes action, arguing that information is the bridge that connects cognitive processes—perception and decision-making—and the physical world, which is manifested in action.[10] Likewise, according to Brown:

> [D]ata is the critical raw material . . . information is its weapon system. . . . A failure to treat data as a strategic resource . . . cedes precious time and space to our adversaries. Like any raw material, data must be harvested, refined, and delivered.[11]

Doing this in a manner that has enduring, effective, and advantageous strategic effect relies on a coordinated strategic manoeuvre approach, and the Russian approach to conflict and competition is perhaps an exemplar of effective strategic level manoeuvre.[12] Russia's New-Generation Warfare (NGW) campaign centres on the 'information struggle', where the battlefield is perception and 'the strategic calculus of the adversary is its centre of gravity'.[13] NGW aims to impose strategic will. The information struggle has several characteristics: it is holistic in that it merges technology and cognitive-psychological attacks aimed at deceiving, disorienting, or demoralising the population or military; it is unified in that it is synchronised across military and non-military activities; it is uninterrupted in that it is conducted throughout the spectrum of war and peace and towards domestic and international audiences.[14] NGW is intended to focus on influencing, shaping, and manipulating perception, decision-making, and behaviour while minimising kinetic engagements.

Western defence forces have generally labelled NGW as hybrid warfare and consider it 'new'. However, as Adamsky has noted, such warfare is not new; it is a part of Russian strategic culture, which favours a holistic approach and considers the 'informational "struggle"' to be the totality of strategic interaction throughout war and peace. 'Competition with the adversary', Adamsky states, 'is seen as protracted, occurring towards, during, and following kinetic phases of interaction'. This is 'somewhat different' from Western military thought, which is more heavily focused on kinetic activity.[15] Furthermore, Western strategic thought considers a 'division of labour' between military and politics, which is not present in a hybrid war.[16]

Defining manoeuvre

A new conception of manoeuvre is useful because it focuses on using the information domain as the foundation or fuel for coordinated action alongside all elements of national power to achieve a position of strategic advantage across the spectrum of conflict and competition, or war and peace. There is a rich discussion in academic literature and military publications about manoeuvre or the 'manoeuvrist approach'. It is generally conceived as the opposite of attrition, which is, in turn, often associated with the positional war experienced during the First World War. Some of the literature draws on Sun Tzu, who noted that the 'acme of skill' was not to win every battle but to 'subdue the enemy without fighting'.[17] In this sense, manoeuvre can also focus on shattering an adversary's will or causing cognitive dislocation that incapacitates or renders the enemy ineffective before a physical confrontation or battle occurs. However, the reality of warfare is that both manoeuvre and attrition have a role to play.

In a Royal United Services Institute article published in 1998, the then Major General John Kiszely provided an excellent overview of manoeuvre.[18] Kiszely's examination, which considered the work of military theorists such as Basil Liddell Hart, J F C Fuller and William Lind, concluded that many debated manoeuvre versus attrition but failed to agree on a definition of manoeuvre. Kiszely discussed Edward Luttwak's use of the term 'manoeuvre warfare', which describes an approach aimed at achieving disruption and exploiting an adversary's weaknesses. Luttwak also referred to 'relational manoeuvre', which focused on 'systematic destruction'—that is, incapacitation and collapse of the enemy's whole system—rather than cumulative destruction arising from a series of attritional engagements.[19]

Clausewitz is also relevant to the discussion of manoeuvre, particularly when considering his concept of 'centre of gravity' or the adversary's source of strength. He believed that to defeat an enemy, one had to overcome the resistance concentrated at its centre of gravity. Accordingly, this centre of gravity needed to be the focus of all efforts to defeat the adversary.[20] Interestingly, Clausewitz's conception of the centre of gravity was revisited by John Boyd, whose work is discussed in US Marine Corp Major Ian T Brown's book *A New Conception of War*. Boyd viewed the centre of gravity 'as those things that permit an organic whole to stay together, whatever they are, moral, mental, physical'. So, when manoeuvre or the manoeuvrist approach is discussed, it is not solely in relation to the fielded military forces. This is particularly the case in the Information Age, where power and pressure can be exerted on various sectors in society, not only on the military. Indeed, in the Information Age, power is not merely measured in military might but also in the ability to influence and sway others to your cause, whatever that may be.[21]

After considering these definitions, there are generally two aspects of manoeuvre that emerge: the use of forces via fire and movement to gain a position of advantage relative to the adversary; and the use of deception, manipulation, 'scheming adroitness', and other approaches that may not involve physical movement. The latter focuses on mentally outmanoeuvring the adversary. This second aspect lends itself most heavily to the Information Age, and it is also why it is necessary to reconsider manoeuvre in addressing 21st-century security challenges. Indeed, Kiszely acknowledges that the concepts of manoeuvre apply at all levels, from the tactical to the strategic. He defined the manoeuvrist approach as 'always seeking alternatives to attrition before resorting to it—attrition being the last resort'.[22]

From this perspective, manoeuvre appears particularly useful when applied at the strategic level. However, although the military instrument was only designed for use in war, it is expected to face numerous security challenges that can exist below the threshold of war, including cyber warfare and counterterrorism. Acknowledging this issue, the Australian Army's current *Land Warfare Doctrine: The Fundamentals of Land Power* defines strategic manoeuvre as

> the coordinated application of the instruments of national power, directly or indirectly, in pursuit of national strategic objectives, and seeks to prevent or contain conflict. The employment of the military element of national power requires strategic manoeuvre to ensure that favourable circumstances are achieved.[23]

This conception of manoeuvre is more appropriate to the Information Age because the lines between war and peace have become quite blurred, and a coordinated approach that uses all instruments of national power—diplomacy, the economy, and the military—is required to address Information Age challenges. When viewed from the strategic perspective, manoeuvre provides a useful foundation for unifying and synchronising the efforts across all the elements of national power to place a nation in a position of advantage relative to an adversary.

Changing the military mind

Basil Liddell Hart suggested that 'the only thing harder than getting a new idea into the military mind is to get an old one out'. The rich discourse on what exactly constitutes a manoeuvrist approach demonstrates its evolution as a concept. Mental models or ways of thinking always change to address novel security challenges, and they are an inherent part of the military profession. The difficulty is changing human attitudes or mindsets to use new ideas that are necessary for tackling novel challenges.

The Information Age has witnessed the rise of non-state actors, cyber threats, and the use of the information environment to exert power via non-military means. Our military and strategic mindsets need to adapt accordingly. Western nations generally tend to use kinetic force to solve problems, but this approach is not always the most appropriate. Manoeuvre concepts offer a holistic approach to meeting these challenges, which have eroded the traditional paradigms of peace and war. As military officers, we need to change our mindset when approaching these security challenges. Just as the ADF once had to think about joint operations, it must now consider how we take real coordinated governmental approaches to our security problems.

Western cultural ideas about the use of military force are clearly defined: the military is for war, and politics is for peace. Yet what lies between these two ends of the spectrum remains unclear; this is why hybrid wars are challenging. Strategic manoeuvre offers a way to address security challenges holistically along the entire spectrum between war and peace.

The allure of battle and the military is 'everything'

In the opening chapters of *Guns of August*, Barbara Tuchman highlights the French obsession with *élan* (enthusiasm, or vigorous spirit) and the philosophy of 'decisive battle' that was a feature of late 19th-century European warfare—the assumption that battles alone could win wars.[24] The problem is that the destruction of enemy forces in a battle or a series of battles was possible in the late 19th century. But in the Industrial Age, where mass production was on an entirely different scale, decisive battles became dangerous as they created a significant risk of protracted war as countries entered conflict on the assumption that it could be won quickly, provided that the enemy was decisively engaged and destroyed in the field.[25] Throwing men against fire over and over again and expecting a different result is the manifestation of insanity—as demonstrated at the battles of the Marne, Passchendaele, and the Somme. Today, in many ways, Western forces have reached a similar turning point. Western forces must choose between sticking to old conceptions of the use of force—of throwing firepower against ideas over and over again, as was done in Vietnam, Afghanistan, Iraq, and Syria, and expecting a different result—or adapting to the Information Age and changing how they think about the use of military force and its limits; and, perhaps, ceding ground to the other elements of national power to achieve policy by other means.

Western obsession with decisive battle remains with us. Nolan provides a comprehensive account of Western military history and the standard resort to decisive battle.[26] The assumption is this: military options equate with rapid action and sometimes absolve national leaders from taking a long-term strategic approach. Undoubtedly, strategy is difficult; it requires what John Lewis Gaddis calls 'historical consciousness' to understand the context of a problem and then plan how to navigate it, with an eye to ensuring future national security.[27] Air power is generally the option of choice in this regard. For instance, when Daesh rose in late 2014, air power was the primary option used to address it. For Australia, this meant joining a coalition of more than seventy nations under the banner of Operation Inherent Resolve.[28] At that point, the reaction was not to sit at a computer and formulate a strategy to provide an enduring solution for fundamentalist terrorism; it was much easier and demonstrated resolve to send some bombers over and 'send them a message', 2000 pounds at a time. However, there is a price to all of this. The allure of battle and sole reliance on military force can lead countries to gamble on the use of force, resulting in those states unknowingly entering a long war of attrition and suffering as a result.[29] Australia's long commitment to the conflicts in Afghanistan and Iraq is a testament to this problem. The Ukraine conflict of 2022 may become another example of this issue.

Civilian leaders generally see the military forces as the first option for almost every security problem; however, an overreliance on military options may lead to the atrophy of other elements of national power and a lack of resourcing. As Rosa Brooks highlights, the military has become 'everything'; it is the Swiss Army knife solution to many security problems, even those for which it is not intended or is not best suited.[30] As there is a clear division of labour between the military and other elements of power in the West, the use of the military as a default option inevitably stretches the boundary between war and peace and makes the challenge of addressing hybrid threats difficult. Only a strategic manoeuvrist approach can appropriately deal with the multifaceted nature of a threat such as global terrorism.

Conclusion

Ideas cannot be bombed to death. They have a viral characteristic, what James Gleick calls 'spreading power', leading to new concepts and, in turn, the evolution and sparking of other thoughts, impressions and developments.[31] A strategic manoeuvre approach that harnesses all the tools at a nation's disposal in a unified and synchronised manner is necessary to fight ideologies. Perhaps it is not only the military mind we need to change; maybe we have a role to play in advising the government about the utility of force in a given strategic situation and offering other solutions to build a long-term solution for security problems. The strategic manoeuvre framework can explain what is termed the whole of government approach.

As this chapter demonstrates, information is a means to unifying and coordinating approaches. Manoeuvre is used to deftly outplay adversaries by engaging in and shaping the information environment in different ways across each of these elements. Nations continuously use diplomatic, trade, and economic actions to develop the information environment. These efforts represent a long-term investment in establishing and maintaining Australia's position within the international community, though a discussion of this position is beyond the scope of this chapter. However, the military element should act as an insurance policy, a symbol of deterrence, rather than the default option when something goes wrong.

Recognising the limits of a military force's effectiveness in meeting 21st-century security challenges such as hybrid warfare is particularly important because air power has been described as 'an unusually seductive form of military strength' that 'appears to offer gratification without commitment'.[32] When coupled with the West's susceptibility to the allure of battle, defence forces become entangled in long wars for which they did not plan, simply because governments invested in the military option and failed to fully explore how it could synchronise other elements of power to address a particular threat or security challenge. The military is not everything, and it is necessary to advise the government and the people that this is the case. The strategic manoeuvre concept can be used as an overarching framework that provides the foundation for developing the discipline to think outside the military and develop enduring solutions to the security challenges in the 21st century.

Notes

1 Zachery Tyson Brown, "Unmasking War's Changing Character," *Modern War Institute*, 12 March 2019, https://mwi.usma.edu/unmasking-wars-changing-character/.
2 Carl von Clausewitz, *On War*, ed. F. N. Maude, trans. J. J. Graham (London, 1909; Project Gutenberg, 2006), www.gutenberg.org/files/1946/1946-h/1946-h.htm.
3 Cathal Nolan, *The Allure of Battle: A History of How Wars Have Been Won and Lost* (New York: Oxford University Press, 2019).
4 For more on the economic basis of the world wars, see: Mark Harrison, ed., The Economics of World War II: Six Great Powers in International Comparison (Cambridge: Cambridge University Press, 1998); Stephen Broadberry and Mark Harrison, "The Economics of World War I: A Comparative Quantitative Analysis" (Unpublished Manuscript, 2 August 2005), http://piketty.pse.ens.fr/files/capitalisback/CountryData/Germany/Other/Pre1950Series/RefsHistoricalGermanAccounts/BroadberryHarrison05.pdf.
5 Alvin Toffler and Heidi Toffler, *War and Anti-War: Survival at the Dawn of the 21st Century* (Boston: Little, Brown, 1993), 58–59.
6 Ibid.
7 See: Roger Hesketh, *Fortitude—The D-Day Deception Campaign* (New York: The Overlook Press, 2000).
8 Carl von Clausewitz, *On War*, trans. Michael Howard and Peter Paret (Princeton: Princeton University Press, 1984), 75.

9 P. W. Singer and Emerson Brooking, *LikeWar: The Weaponization of Social Media* (Boston: Houghton Mifflin Harcourt, 2018), 4.

10 Marc Romanych, "A Theory-Based View of IO," *IO Sphere,* Spring 2005: fig. 1.

11 Brown, "Unmasking War's Changing Character."

12 Such an approach, while effective in Crimea and the Donbas, was not followed with any coordinated zeal in the 2022 Crimea invasion. Russia's subsequent difficulty in subduing Ukrainian forces reinforces the broader arguments of this chapter.

13 Dmitry Adamsky, "From Moscow with Coercion: Russian Deterrence Theory and Strategic Culture," *Journal of Strategic Studies* 41, no. 1–2 (2018): 54.

14 Ibid., 42.

15 Ibid., 52.

16 See, also: Thomas Rid and Marc Hecker, *War 2.0: Irregular Warfare in the Information Age* (Westport: Praeger Security International, 2009).

17 Sun Tzu, "The Art of War," in *Oxford Essential Quotations*, ed. Susan Ratcliffe, 5th ed. (n.p.: Oxford University Press, 2017), www.oxfordreference.com/view/10.1093/acref/9780191843730.001.0001/q-oro-ed5-00010536?rskey=nM61bL&result=3559.

18 John Kiszely, "The Meaning of Manoeuvre," *The RUSI Journal* 136, no. 6 (1988): 36–40.

19 Kiszely, "The Meaning of Manoeuvre," 37.

20 Quoted in Ian Brown, "John Boyd on Clausewitz: Don't Fall in Love with Your Mental Model," *The Strategy Bridge* Website, 22 March 2018, https://thestrategybridge.org/the-bridge/2018/3/22/john-boyd-on-clausewitz-dont-fall-in-love-with-your-mental-model.

21 Ian Brown, *A New Conception of War: John Boyd, the Marine Corps, and Maneuver Warfare* (Quantico: Marine Corps University Press, 2018), 212–13.

22 Kiszely, "The Meaning of Manoeuvre," 39.

23 Australian Army, *Land Warfare Doctrine 1: The Fundamentals of Land Power–2017* (Canberra: Commonwealth of Australia, 2017), 32.

24 Barbara Tuchman, *The Guns of August* (New York: Presidio Press, 2004), 39–41.

25 Lawrence Freedman, *The Future of War: A History* (New York: Hachette, 2017), 10.

26 See: Nolan, *The Allure of Battle.*

27 John Lewis Gaddis, *The Landscape of History: How Historians Map the Past* (Oxford: Oxford University Press, 2002), 8–11.

28 Renee Westra, *Syria: Australian Military Operations*, Parliamentary Library Research Paper Series 2017–2018 (Canberra: Commonwealth of Australia, 20 September 2017), https://parlinfo.aph.gov.au/parlInfo/download/library/prspub/5526262/upload_binary/5526262.pdf.

29 Freedman, *The Future of War*, 10.

30 Rosa Brooks, *How Everything Became War and the Military Became Everything: Tales from the Pentagon* (New York: Simon & Schuster, 2016).

31 James Gleick, *The Information: A History, A Theory, A Flood* (New York: Pantheon Books, 2011), 310–11.

32 Eliot Cohen, "The Mystique of US Air Power," *Foreign Affairs* 73, no. 1 (January 1994): 109.

References

Government sources

Australian Army. *Land Warfare Doctrine 1: The Fundamentals of Land Power–2017.* Canberra: Commonwealth of Australia, 2017. https://researchcentre.army.gov.au/sites/default/files/lwd_1_the_fundamentals_of_land_power_full_july_2017.pdf.

Westra, Renee. *Syria: Australian military operations*. Parliamentary Library Research Paper Series 2017–2018. Canberra: Commonwealth of Australia, 20 September 2017. https://parlinfo.aph.gov.au/parlInfo/download/library/prspub/5526262/upload_binary/5526262.pdf.

Published sources

Adamsky, Dmitry. "From Moscow with Coercion: Russian Deterrence Theory and Strategic Culture." *Journal of Strategic Studies* 41, no. 1–2 (2018): 33–60.
Broadberry, Stephen, and Mark Harrison. "The Economics of World War I: A Comparative Quantitative Analysis." Unpublished manuscript, 2 August 2005. Pdf file. http://piketty.pse.ens.fr/files/capitalisback/CountryData/Germany/Other/Pre1950Series/RefsHistoricalGermanAccounts/BroadberryHarrison05.pdf.
Brooks, Rosa. *How Everything Became War and the Military Became Everything: Tales from the Pentagon*. New York: Simon & Schuster, 2016.
Brown, Ian. *A New Conception of War: John Boyd, the Marine Corps, and Maneuver Warfare*. Quantico: Marine Corps University Press, 2018.
Clausewitz, Carl von. *On War*. Edited by F. N. Maude. Translated by J. J. Graham. London, 1909; Project Gutenberg, 2006. www.gutenberg.org/files/1946/1946-h/1946-h.htm.
———. *On War*. Translated by Michael Howard and Peter Paret. Princeton: Princeton University Press, 1984.
Cohen, Eliot. "The Mystique of US Air Power." *Foreign Affairs* 73, no. 1 (January 1994): 109–24.
Freedman, Lawrence. *The Future of War: A History*. New York: Hachette, 2017.
Gaddis, John Lewis. *The Landscape of History: How Historians Map the Past*. Oxford: Oxford University Press, 2002.
Gleick, James. *The Information: A History, A Theory, A Flood*. New York: Pantheon Books, 2011.
Harrison, Mark, ed. *The Economics of World War II: Six Great Powers in International Comparison*. Cambridge: Cambridge University Press, 1998.
Hesketh, Roger. *Fortitude–The D-Day Deception Campaign*. New York: The Overlook Press, 2000.
Kiszely, John. "The Meaning of Manoeuvre." *The RUSI Journal* 136, no. 6 (1988): 36–40.
Nolan, Cathal. *The Allure of Battle: A History of How Wars Have Been Won and Lost*. New York: Oxford University Press, 2019.
Rid, Thomas, and Marc Hecker. *War 2.0: Irregular Warfare in the Information Age*. Westport: Praeger Security International, 2009.
Romanych, Marc. "A Theory-based View of IO." *IO Sphere* (Spring 2005): 14–18.
Singer, P. W., and Emerson Brooking. *LikeWar: The Weaponization of Social Media*. Boston: Houghton Mifflin Harcourt, 2018.
Sun, Tzu. "The Art of War." In *Oxford Essential Quotations*, edited by Susan Ratcliffe. 5th ed. n.p.: Oxford University Press, 2017. Oxford Reference.
Toffler, Alvin, and Heidi Toffler. *War and Anti-War: Survival at the Dawn of the 21st Century*. Boston: Little, Brown, 1993.
Tuchman, Barbara. *The Guns of August*. New York: Presidio Press, 2004.

15 An overview of Australian space power, from desert rockets to new beginnings

Amy Hestermann-Crane

Australia has been involved in space since the 1940s, boasting successes and facing challenges of capability and policy to this day. From the beginning of this involvement, space has been a military domain used to further national security objectives. Yet this military expansion has relied on the cooperation of military and civilian sectors to benefit from technical expertise and avoid duplication.[1] This cooperation increasingly includes commercial sectors, with organisations such as Rocket Lab, SpaceX, and One Space aiding in launching military and commercial satellites.[2] Australia's space history is no exception, with the nation's involvement in space characterised by both international cooperation and the blending of military and civilian elements.

The momentum of Australia's early participation in the space domain did not last; several decades of reduced governmental support and sovereign research and development ensured that after the 1960s, Australia faded as a space-faring nation.[3] Yet the turn of the 21st century saw renewed interest in space, rapidly expanding in the 2010s to the present day. Todd Harrison notes that militaries worldwide, including the Australian Defence Force (ADF), are 'integrating advanced space systems into military operations in increasingly sophisticated ways'.[4] The global community has integrated space systems into multiple facets of everyday life, security, and emergency services, leading to the domain's evolution into a global commons.[5] This reliance on space makes understanding the current space environment, including associated risks and the domain's potential future, essential to ensuring the continued use of this vital domain.

This chapter will examine the historical use of the Australian space domain, providing a foundation for the developments within this realm, the Australian Government's motivations, and how the latter shaped these developments. Building on this discussion, it will move into the contemporary domain, considering how the ADF and civilian organisations use space to further military objectives. With space being a realm of technological advancement and continuing military and commercial benefits, the future of space is filled with both possibilities and hurdles. This chapter considers Australia's future in space and examines space policy requirements, counter-space capabilities, and the benefits of blended cooperation and sovereign capability to best secure the ADF's access to this vital domain.

DOI: 10.4324/9781003230656-20

Historical use of Australian space

In contrast to other nations, Australia's space history is not the subject of a large amount of literature. Yet the history of Australian use of space is important in understanding how the nation arrived at its contemporary place in the space domain. It is tempting to see Australia as an early leader in space, embracing the emerging domain enthusiastically as it actively participated from the 1940s into the 1960s. Indeed, Australia was a leading nation in the foundation of the United Nations Committee on the Peaceful Uses of Outer Space (UNCOPUOS) and became the fourth 'space-faring' nation with the launch of its first satellite in 1967.[6] However, as Tristan Moss demonstrates through a review of government policy and decision-making choices, the reality is quite different.[7] Even a short overview of Australia's space domain history illustrates that Australia has primarily been a reactive participant, pushed by its desire to strengthen relationships with historic and emerging allies, as well as the benefits these relationships and potential space capabilities could have for national security.[8] These drivers have shaped the aspirations and commitments to space activities that successive Australian Governments have sought to achieve and have made between the 1940s and the 1990s.

Although Australia's interest in human exploration and exploitation of space began in the mid-1940s, it would be remiss to state that this represented the nation's first interest in the space domain. Like many cultures, Aboriginal and Torres Strait Islander peoples have used the Australian sky in multiple facets of their lives.[9] Stellar Scintillation, or the 'twinkling of stars', held practical applications for the Torres Strait Islanders in Australia's northern regions, where it was used to predict weather and seasonal changes.[10] Similarly, the orientation of the Celestial Emu constellation aided the Wiradjuri, Kamilaroi, and Euahlayi peoples of central-west New South Wales in wildlife pattern interpretation, food harvesting, and animal conservation.[11] With increasing academic interest in Indigenous astronomy, the complex means by which the First Nations peoples used the night sky are increasingly recognised. Yet despite the long tradition and importance of space to First Nations peoples, their interests were unfortunately not considered during Australia's early steps into space research and development in the 1940s.

Instead, Australia's first steps into space evolved from its alliance with the United Kingdom. As the United Kingdom entered the 'Missile Age', its small geography meant any new rocket range would need to be overseas since existing ranges, both at home and abroad, were too short or required firing into the sea. The United Kingdom, therefore, required a new location. Australia fit the bill, as security requirements and weapon range distance put it ahead of other Commonwealth states, and disadvantages could be relatively easily overcome.[12] The Weapons Research Establishment (WRE) at Woomera, South Australia, was subsequently established in 1946 and declared under the Defence Force Regulation 35 in 1947.[13] The site remained under the Defence group administration through the Department of Supply.

Australian Government support for the WRE came from its close relationship with the United Kingdom and enduring policy of investing in overall Commonwealth Defence capability after the Second World War.[14] Now known as RAAF Woomera Range Complex, the WRE was initially a highly classified joint British-Australian venture that sought to develop and test missiles and 'war materials' in a vastly unpopulated remote location.[15] By the 1950s, the WRE had the infrastructure for inter-continental ballistic missile (ICBM) testing, a missile that reaches space during flight.[16] Although initially established with all three Services of the ADF providing personnel for various functions, by 1955, most military personnel were replaced with civilian staff. While the International Geophysical Year of 1957 is famously remembered for the launch of the Soviet Union's Sputnik satellite, Australia and Britain were also among the seventy nations that contributed to this global scientific research effort.[17] Australia and Britain launched sounding rockets from Woomera, a valuable atmospheric research tool that was considerably cheaper and simpler than contemporary satellites.[18]

By the end of the 1950s, the United States also began to take an interest in Woomera. The United States had already, by this stage, firmly established its position as a superpower and an emerging Australian ally following the Second World War, but it now began to show interest in Australia's geographic location relevant to space.[19] In 1956, through the Smithsonian Institute, the United States requested the Australian Academy of Science assist with tracking American rockets and satellite launches. Although the United States stressed the scientific benefits of the venture, it was the Defence Committee that was asked for comment, as the first station was likely to be built at Woomera. The Committee recommended the Australian Government to approve the tracking station due to the 'military significance' if the United States found success in its satellite programme. The station became operational in October 1957, shortly before Sputnik was launched, and supported the launch of Explorer 1, the United States' first satellite, in 1958.[20] A second US ground station was built at Nurrungar near Woomera in 1969, and the station continued to support the US Defence Surveillance Program until its closure in 1999.[21]

However, the 1960s witnessed a significant change to Woomera and the Defence monopoly at the site following the UK Government's decision to cancel its Woomera-based Blue Streak ICBM programme in 1959.[22] To recuperate financial losses, the UK Government sought European and Commonwealth support for developing the Blue Streak rocket for civilian satellite launches.[23] Australia did not support the shift away from security objectives; the government considered water conservation, education, internal communications, and transport to be higher priorities than the 'national prestige' of space use. Nevertheless, Australia became a partner of the European Launcher Development Organisation (ELDO) alongside Belgium, France, Germany, Italy, the Netherlands, and the UK, in April 1962.[24] With Australia providing Woomera as the testing facility, ELDO developed the Europa-1 rocket, which undertook ten launches, five of which failed during the programme life of 1964–70.[25] In 1967, the ELDO programme entered a joint venture with the WRE, the University of Adelaide, the

US Department of Defense, the National Aeronautics and Space Administration (NASA), and the UK Ministry of Technology to design Australia's first satellite. The WRE satellite was designed to increase understanding of upper atmosphere effects on weather and climate and aid the United States in gaining physical data for research programmes.[26] However, the international space programme proved short-lived. ELDO moved the programme to Kourou, French Guiana, in 1970, and the Australian Government declined to remain within the organisation once it left Australian soil. The United Kingdom also withdrew from ELDO in 1971 to join the US space programme. After nine years of working on Europa rockets, ELDO formally closed its space launcher programme in 1973.[27]

The late 1950s and 1960s saw Australian national space policy driven by international engagements and alliances with the United Kingdom and the United States.[28] Australia was active in the foundation of UNCOPUOS in 1958, and this involvement in part contributed to the Australian Government's realisation that it required an interdepartmental committee to advise on space-related matters.[29] The Department of Supply and the Australian Government's scientific research agency, the Commonwealth Scientific and Industrial Research Organisation, held primary responsibility for Australian space activities in the 1960s. Each organisation took control of Defence and civilian space endeavours, respectively. However, the government continued to remain resistant to either a civilian or independent (versus cooperative) space programme.[30]

While the Australian Government remained less favourable to civilian space programmes in the 1960s, it still supported projects that involved national security and alliance outcomes. Australia signed the Joint Defence Facility Pine Gap agreement in 1966, and the facility opened in 1970. At the time, the Australian Government promoted the military and scientific research benefits of the facility; however, it did not reveal that the project supported the United States' signals intelligence collection satellites.[31] Likewise, the Australian Government valued its continuing relationship with the United States and supported US ground tracking stations in Australia. NASA, however, determined that Woomera was not beneficial for deep space communications and moved their complex to a new location. In the early 1960s, Tidbinbilla valley was chosen for its proximity to Canberra and radio interference-shielded location. Thus, from the Canberra Deep Space Communication Complex, Australia aided the Apollo 11 moon landing.[32]

The following decade saw little innovation for Australian space activities. Again, it was international interests that engaged Australia in space activity. In the 1970s, the United States launched Skylab, a project that sought to assess the social, physiological, and practical elements of human survival in orbit. When the Saturn-V rocket launched the orbital workshop in 1973, two Australian-based tracking stations supported Carnarvon in Western Australia and Honeysuckle Creek in the Australian Capital Territory. Skylab later caused a commotion when it re-entered Earth's atmosphere in the early hours of 12 July 1979, with debris falling over a wide area in Western Australia, from Perth to Kalgoorlie and the Shire of Esperance.[33] Forty years later, one eyewitness recalled that 'it was the best fireworks display you would ever see' as he recounted watching the debris 'rocketing' across the sky.[34] At the time, the Shire of Esperance issued NASA with the now-famous littering fine, which, although a joke, raised awareness on how space and space-faring nations could affect others.[35]

The 1980s and 1990s saw the Australian space industry fail to launch. Brett Biddington highlights numerous failed attempts to establish commercial launch facilities and a failure to establish commercial satellite operations.[36] The Australian Government reacted to Sir Russell Madigan's 1985 report, which supported establishing government structures for Australian space management; however, the Australian Space Office and the Australian Space Board remained under-funded non-statutory bodies and were disbanded shortly after in the mid-1990s by the Howard government.[37] Although the *Space Activities Act* was created in 1998 to manage space activity liability, it would not be until the mid-2000s that Australian interest in space began to re-emerge.[38]

Australia's contemporary space domain

In Australia's contemporary space domain, the ADF continues the tradition of integrating military and civilian participation to further objectives in security, civilian, and commercial realms. It is no secret the near-Earth space environment is increasingly becoming contested, congested, and competitive. As the ADF embraces space and its competitive advantage, it also recognises the increasing risks to unrestricted access, and it is taking proactive steps to protect Australian exploitation of this domain.[39]

Space recaptured the Australian Government's imagination in the late 2000s. Fortunately, Australia held—and still holds—numerous benefits to space domain research and development. Biddington highlights several enduring factors: few geographic limitations, radio quietness due to vast unpopulated areas free of radio interference noise, favourable longitude for ground stations to aid in continuous coverage of satellite tracking, and a latitude that provides a unique perspective on space observation and activities.[40] These elements continue to drive interest in Australia as a location for space, and this international interest is mirrored domestically. Recent decades have seen rapid increases in government policy and Defence investment in space capabilities.

At the turn of the century, there was slow progress in the space sector. Having disbanded the Australian Space Office and the Australian Space Board a decade earlier, the Howard government reinforced its position on space in its *Space Engagement Statement* of 2003, which stated that there was no need for Australia to pursue space self-sufficiency.[41] However, the Australian attitude towards space began to change following the 2007 Australian federal election, in which the Labor Party was victorious. In 2008, the Senate Standing Committee on Economics examined options to strengthen and expand Australia's position concerning space science and industry. The subsequent report was released shortly before the Global Financial Crisis (GFC) affected numerous national economies worldwide.[42] The Committee's report was followed soon after by the *2009 Defence White Paper*, the first to mention space in any significance in eight years.[43] The

2009–10 Australian federal budget also allocated \$40 million to establish an Australian Space Research Program, and additional funds for a Space Policy Unit (SPU) were provided as part of the Rudd government's stimulatory response to the GFC.[44] The SPU was tasked with creating a national civil space policy—eventually titled *Australia's Satellite Utilisation Policy*—which it presented to Cabinet in 2013. However, no funds were allocated to the direct realisation of this policy at the time.[45]

Several years later, in 2017, space again emerged at the forefront of government policy when the Australian Government announced it would establish the Australian Space Agency (ASA), with the organisation set to open in 2018.[46] The ASA was principally tasked to create a 'globally respected Australian space industry that lifts the broader economy, inspires and improves the lives of Australians—underpinned by strong international and national engagement'.[47] With Australia holding only a passing interest in this quickly evolving domain over the previous decades, the Australian Government grew concerned that domestic space policy and legislation no longer held relevancy. Accordingly, the agency was also tasked with reviewing domestic space legislation, specifically the *Space Activities Act 1998* (Cth), which was by this time out of date. This review had two outcomes. First, the *Space Activities Act 1998* (Cth) was amended on 31 August 2018. This amendment was followed by the *Space (Launches and Returns) Act 2018* (Cth), which came into force the following day and made several significant changes to the previous act.[48] These Acts would become central in determining how Australian and foreign entities participate in space and space re-entry.[49] Despite a lack of resources, as it grows, the ASA could be well suited to advancing Australia's space diplomacy efforts and allowing the nation to contribute to shaping this domain.[50]

In terms of space security, Australia has continued its tradition of international partnership. The 2010 Australia–United States Ministerial Consultations (AUSMIN) Space Situational Awareness Partnership saw space mentioned in these discussions for the first time, with concerns about access to space and the risks of counter-space technologies highlighted specifically.[51] This AUSMIN forum also recognised the 'crucial nature of satellites' and the importance of space security.[52] More recently, the 2021 AUSMIN forum specifically mentioned the importance for both nations to establish 'shared capabilities in Space Domain Awareness, Space Command and Control, Satellite Communications, and Positioning, Navigation, and Timing'.[53] The trilateral Australia–United Kingdom–United States security pact, announced the same day as the 2021 AUSMIN, further strengthened these space-related objectives.[54] These pacts and agreements echo Australia's early involvement in space, again tying Australian space capabilities and national security to those of our traditional partners.

The ADF has embraced the space domain with vigour in recent years, as developments in Defence policy and significant interest in space capabilities focus on military objectives. The Australian Government has agreed to invest approximately \$7 billion into the space domain over the next decade to meet these interests. This investment highlights the government's acknowledgement that 'access to space is critical to ADF warfighting effectiveness, situational awareness, and the delivery of real-time communications and information'.[55] As Malcolm Davis suggests, the advancement in language from the *2016 Defence White Paper* illustrates Defence's acknowledgement that space capability access cannot be guaranteed in a contested environment.[56]

In 2021, the ADF formally announced the establishment of a joint space command, with then Chief of Air Force Air Marshal Mel Hupfeld appointed as the space domain lead and Air Vice-Marshal Catherine Roberts as the first Head of Space Division.[57] Air Marshal Hupfeld subsequently led a Space Domain Review, set for completion by the end of 2022, which presents a myriad of exciting possibilities relating to space capability and focus. With this announcement, Australia joined the ranks of nations with a space command, including Canada, France, Germany, India, Japan, and the UK. Similarly, China, Russia, and the United States have a space force or equivalent as a separate military arm responsible for space capabilities.[58]

The ADF has, however, been using space capabilities in one form or another long before the formation of a dedicated space command was announced. Space units within the ADF already exist, including the Australian Space Operations Centre (AUSSpOC) and No. 1 Remote Sensor Unit (1RSU). The AUSSpOC, although smaller in size, appears to mirror similar Allied space operations centres such as the US Joint Space Operations Command.[59] This unit has two primary goals: to provide space domain awareness (SDA) and space control—essentially all the measures taken by the ADF to use space.[60] Also, 1RSU operates Australia's long-range radars, is responsible for operating the C-Band space surveillance radar and space surveillance telescope, and supports the ADF Space Situational Awareness (SSA) mission.[61] Royal Australian Air Force (RAAF) units such as the AUSSpOC and 1RSU are critical to the ADF more broadly, as they provide analytic advice on the effects other space users' actions may have on the ADF and its operations.[62]

The 2010s also saw the ADF return to satellite launches, performed through international launch programmes. The Buccaneer Risk Mitigation Mission cube satellite (CubeSat), a joint project between Defence Science and Technology (DST) and UNSW Canberra, was launched from Vandenberg Air Force Base, California, in November 2017. Buccaneer is a small satellite, measuring as a 3U CubeSat with a mass of up to four kilograms. Its mission is to use the high-frequency receiver payload to provide performance calibration support to Australia's Jindalee Over-the-Horizon Radar Network and test novel DST design elements, such as the cross-antenna-release design.[63] Deemed a success when the satellite launched, the project was an important step toward the RAAF-funded M1 mission.

The RAAF-commissioned M1 mission was launched aboard a SpaceX Falcon 9 in a rideshare payload in December 2018. The M1 represents the first in a three-mission collaboration effort between RAAF and UNSW Canberra, designed to 'push the boundaries of small space technology' for the ADF and the broader Australian space sector.[64] The Space Environment Research Centre developed one such technology, Research Program 4, a payload programme that aims to demonstrate a 'practical active collision avoidance system using photo pressure'. Unfortunately, contact could not be made with M1 once the satellite achieved orbit.[65] There was greater success in 2020 when UNSW Canberra achieved communication with the M2 Pathfinder in June. The small satellite demonstrated Australia's space development capability, being designed, assembled, and tested within ten months. The mission for M2 Pathfinder was described as providing a 'unique opportunity to support Australia's defence and national security capabilities'.[66]

It is clear Australia benefits from space-based intelligence, surveillance, and reconnaissance (ISR) capabilities, having launched demonstrator satellites in line with many other nations.[67] However, ISR assets are typically classified, making it notoriously difficult to confirm their existence or capabilities. As such, this chapter will not attempt to provide a detailed scope on potential space-based ISR. That said, it is possible to discuss how space capabilities support Australia's humanitarian aid and disaster relief efforts. Space capabilities are frequently used to help identify and prevent natural disasters and support rescue and recovery efforts.[68] For example, as part of a combined international effort, the ADF launched Operation Bushfire Assist in the summer of 2019–20 to aid the nationwide effort to mitigate and combat bushfire damage. Tangible efforts involved terrestrial support of more than 6500 personnel in clearance, repairs and fire combat, maritime and air logistics support, and opening bases to support emergency service personnel and evacuees.[69] However, these efforts were also supported by earth observation satellites used to track the ever-growing fires and aid in firefighting efforts, a common practice during natural disasters.[70]

Satellite communication (SATCOM) is the primary method of long-range, beyond line-of-sight communication for the ADF.[71] The ADF's indigenous SATCOM satellite is Optus C1, launched in 2003 from French Guinea. The largest dual-use communication satellite when it was launched, it provides military communication coverage over Australia, New Zealand, nearby offshore islands, Papua New Guinea, Hawaii, and Southeast Asia alongside television and Aurora radio and television services to remote Australia.[72] The government extended the SATCOM contract with Optus 'well into the next decade' in 2017, with the assurance that Optus will maintain the operational capability of C1 until at least 2027.[73] The ADF also leverages a global network of SATCOM satellites for operations abroad, and organisations such as the DST group likewise support the ADF's SATCOM capability.[74] The Cortex satellite spectrum monitoring system was first trialled in the lead up to Exercise Talisman Sabre in 2013 and is currently being further developed under a three-year bilateral collaborative research agreement awarded in 2019. This prototype system has been designed as a monitoring and management system to detect anomalies within, and improve situational awareness of, Defence satellite networks.[75]

SSA is crucial to the protection of space-based assets and terrestrial-based lives and property, as the Skylab re-entry demonstrated. Currently, Australia holds no sovereign SDA capability, relying solely on the US.[76] Australia is partnering with allied militaries, industry, and Australian universities to advance its space surveillance capabilities and aid sovereign and global SSA. It is critical that Australia has a sovereign SDA capability before establishing an indigenous satellite pro-gramme.[77] To this end, the RAAF acquired a space surveillance telescope in Exmouth, Western Australia, through a partnership with the United States Space Force. The telescope will integrate into a global space surveillance network, with a mission to provide SDA for Australia and the US.[78] Training RAAF person-nel to use this telescope has already begun in preparation for when the telescope becomes operational in 2022, and the government is also leveraging expertise within industry and academia to develop sovereign capability.[79] Through a memo-randum of understanding signed in July 2021, the University of Tasmania will combine its expertise in space observation tracking with the space radar technol-ogy and system development of HENSOLDT (a multinational aerospace com-pany) to create the 'Southern Guardian' SDA system.[80]

Australia continues to engage in international space engagement opportunities. The vast unpopulated areas in central Australia remain a desirable location for space re-entry and recovery operations. In 2020, Japan Aerospace Exploration Agency received permission for its Hayabusa 2 mission capsule to re-enter and land at RAAF Base Woomera. The capsule contained samples of the near-Earth asteroid Ryugu, which would allow scientists to examine material believed to originate from the formation of the solar system.[81] Although this was a purely scientific mission, it involved the ADF. The RAAF was responsible for the base and its impact areas, highlighting another means of Defence collaboration within the broader space domain.

Many areas of space capability are currently being improved within Australia, such as sovereign launch facilities. Although a space launch vehicle has not been launched from Australian territory for decades, companies have fired rockets in recent years. Southern Launch succeeded in launching a space-capable rocket 85 kilometres high in 2020.[82] Unfortunately, Southern Launch's second proof of launch capability had to be cancelled, as the Taiwanese rocket caught fire during its third launch attempt in September 2021.[83] On the other hand, Black Sky Aero-space successfully launched their Australian-designed and built rocket in Novem-ber 2021. The rocket is still in the test and development phase, reaching an apogee of approximately nine kilometres, but it represents an important step towards a future Australian sovereign launch capability.[84]

Looking forward: future space possibilities

Several areas within the space domain would benefit from continued investment to properly establish. Notably, Australia lacks a national space policy that unifies the defence, civilian, and commercial sectors.[85] The seeds of sovereign capability have already been planted, with projects such as Southern Guardian providing a means for Australia to maintain capability in a contested environment without wholly relying on its international partners. However, Australia does not have any means of counter-space capability. Finally, space education and growing sover-eign capacity to excite and train the next generation is a continued area that could significantly benefit the nation.

Australia, along with the international community, increasingly recognises the ways in which space is integral for national security and economic prosperity, domestically and abroad. In 2020, the Department of Home Affairs classified space systems as 'critical infrastructure' needing protection; however, the effects of this recognition are yet to be reflected in government policy.[86] Through such recognition, Australia must work towards developing space policy and building its diplomatic space efforts. The international community also recognises the need for improved space policy, particularly due to the growing possibility of conflict within the space domain. To this end, experts from around the world are collaborating on the Woomera Manual on the International Law of Military Space Activities and Operations[87] (Woomera Manual).[88] Currently, space activities are to adhere with international law, including any potential conflicts that were to arise. Yet space has unique characteristics from terrestrial warfighting domains, making certain legal aspects unclear when applied to military space activities. The Woomera manual attempts to provide clarity on these military activities and operations, while respecting the 1967 Outer Space Treaty principle of 'peaceful purposes'.[89] Cassandra Steer, an Associate Expert for the Woomera Manual, states the Woomera Manual's role is 'to limit, rather than legitimise space warfare'.[90]

Air Marshal Hupfeld (Ret.) also acknowledges that space is an operational domain, where Australia will adhere to responsible space behaviours and not 'weaponise space'.[91] However, other militaries do not share the same intention. Military doctrine and space policy language have become increasingly aggressive, with nations such as China and the United States reverting from the historical position of keeping space stable to the 'weaponisation' of space.[92] The term 'weaponisation' has come to encompass the development and use of kinetic anti-satellite (ASAT) weapons, either direct-ascent missiles or on-orbit technology.[93] These ASAT tests, most recently conducted by Russia, are irreversible and show the serious risk to the security and stability of assets in near-Earth orbits.[94] Yet it is important to note that Russia is only the latest nation to conduct kinetic ASAT testing; China, India, and the United States have conducted tests in the previous decade.[95] These tests pose risks to other satellites, including the International Space Station, with the potential to trigger the Kessler syndrome, a cascading effect of ever-increasing debris in low Earth orbit that leads to widespread destruction complicating subsequent human use of space.[96]

With debris from these tests lingering in space indefinitely, the risk they pose will remain for decades after the event. As space becomes increasingly congested, the risk to active space assets also commensurately increases. This threat posed by space debris highlights the benefits of developing means of cleaning the near-earth space environment, both from small and large pieces of debris. These efforts began with the recent launch of the ELSA-d by Japan-based Astroscale in March 2021. The mission is to demonstrate a viable means of clearing some of the 8000 metric tons of space debris, focusing on debris originating from future launches rather than current defunct satellites.[97] The benefits for Australia in developing a viable clean-up solution go beyond potential economic or prestige benefits. Not only will it reduce the risk of the Kessler syndrome occurring, but it would also aid in providing continued access to the satellites on which Australia relies for security and critical infrastructure. Moreover, removing defunct satellites will allow improved satellites to launch for continued development in space capabilities.

The protection of space assets goes beyond their physical means. As with all military domains, there are counter techniques to protect one's own capabilities while denying those of an adversary. Nations such as China, Russia, and the United States have already developed numerous means of counter-space capabilities, both destructive and non-destructive.[98] Kinetic ASATs, a physically destructive capability, are only one means of counter-space technology. Non-kinetic means encompass both destructive and non-destructive. Directed energy weapons can be either non-destructive or destructive, depending on their energy output. For instance, high-powered lasers can temporarily 'dazzle' or permanently blind/deafen satellites.[99] Conversely, electronic warfare methods of jamming or spoofing are considered solely non-destructive means of counter-space technology.[100] In recognition of the critical need to protect sovereign space assets, the Australian Minister for Defence, Peter Dutton, announced the establishment of Defence Project 9358 in July 2021. This project is designed to explore options for the ADF to acquire 'ground-based Space Electronic Warfare capability'. Such a capability is non-destructive (meaning no damage to the space environment) and intended to counter space threats, thereby assuring continued access to Australia's space-based communications and ISR.[101] The requirement for Australia to possess future counter-space technology is arguably highly likely as space assets continue to improve in the ISR realm. The current alignment to non-destructive methods of protection is consistent with Defence's undertaking to refrain from weaponising space and adding to the debris layer within the domain.

For military cooperation, there is greater excitement than merely developing alliances with our traditional partners. With space being a global domain and one whose effects span international borders, it is not surprising that partnerships are being increasingly developed. The Combined Space Operation (often referred to simply as CSpO) initiative has formed and is the basis for broadening cooperation in the space domain.[102] France and Germany were welcomed into the initiative in February 2020, alongside Australia, Canada, New Zealand, and the United Kingdom, with the possibility of Japan entering the fold once their space command is established.[103] It would be beneficial to direct resources to both civilian and military diplomacy efforts to help Australia shape and influence the international dialogue on space, a vital area of national interest.[104]

Industry and scientific research developments will also continue to aid in Defence. As natural disasters become increasingly frequent and devastating, earth observation satellites will continue to be a vital resource in preventing or identifying these situations. The ASA has released its roadmap for 2021–30, which includes the need for Australia to establish an indigenous earth observation capability for natural disaster mitigation and management.[105] Fireball, a Queensland-based company, announced the development of a purpose-built satellite to detect Australian bushfires within minutes of ignition. This satellite is set to launch in early 2022. It will aid firefighting services in identifying, locating, and appropriately resourcing efforts across the country—a capability that will be particularly beneficial for remote locations where identification delays can cause serious damage.[106] This satellite could also aid the ADF when used in Australian humanitarian relief efforts, perhaps even mitigating the requirement for ADF support to emergency service efforts.

Conclusion

Australia has a mixed history with space. Despite early participation in leading technologies, an undercurrent of resistance stunted Australian space capability towards the end of the 20th century. As the representative of a small nation, the Australian Government understandably placed terrestrial concerns of the people above a capability that was poorly understood and seen not to hold tangible benefits to society. Yet as space constitutes an ever-increasing presence and importance to military capability, economic stability, and quality of life, this domain is now integral to continued operations and societal functions. To its credit, as the 21st century steadily marches forward, so too does Australia's interest and participation in the space domain. The Australian Government and the ADF have realised the key role space plays in Australian society and national security efforts. They have demonstrated this change in attitude in policymaking and varied investment within the domain through multiple vectors.

As a medium power, Australia will continue to rely on its international partners for security and capability. With the current government's continued investment and interest in both Defence and industry, Australia is beginning to form a sovereign capability to contribute domestically and to its partners in various global space efforts. Australia must broaden and deepen its space diplomacy relationships to provide increased security and access to capability. Further, active participation in diplomatic efforts will allow Australia to influence the development of governance frameworks and norms relating to space in international fora.

The ADF has already taken significant steps in ensuring space security in recent years. The establishment of a whole-of-Defence space command will allow for increased expertise and focus on military capability within the domain. Continuing traditional partnerships has enabled Australia to place itself as a partner of choice, bringing space capability to the region. Australia's vast unpopulated areas, relative radio quietness, and beneficial geography in contributing to global space tracking coverage are all advantageous to various space-related operations. Maximising these benefits provides Australia with multiple avenues to becoming a contributing partner in building global space relationships. These benefits will only increase as the ASA develops Australia's space industry.

Notes

1 Elizabeth S. Waldrop, "Integration of Military and Civilian Space Assets: Legal and National Security Implications," *Air Force Law Review* 55 (2004): 159.

2 Vi Tran, "Rocketlab to Carry Australian Payloads to Low-Earth Orbit," *Space Australia*, 19 February 2021, https://spaceaustralia.com/news/rocketlab-carry-australian-payloads-low-earth-orbit; Danielle Sempsrott, "NASA Announces Date for SpaceX's 24th Cargo Resupply Mission," *NASA Blogs* (blog), 24 November 2021, https://blogs.nasa.gov/kennedy/2021/11/24/nasa-announces-date-for-spacexs-24th-cargo-resupply-mission/; Andrew Jones, "Chinese Companies OneSpace and iSpace Are Preparing for First Orbital Launches," *SpaceNews,* 24 January 2019, https://spacenews.com/chinese-companies-onespace-and-ispace-are-preparing-for-first-orbital-launches/.

3 Tristan Moss, "'There Are Many Other Things More Important to us Than Space Research': The Australian Government and the Dawn of the Space Age 1956–62," *Australian Historical Studies* 51, no. 4 (2020): 442–58; Tristan Moss, *Unifying Space: Australia Needs a Whole of Government Space Policy*, Black Swan Strategy Paper Series (Crawley, Western Australia: University of Western Australia Defence and Security Institute, 2020), 7–9; Kerrie Dougherty, "Lost in Space: Australia Dwindled from Space Leader to Also Ran in 50 Years," *The Conversation*, 22 September 2017, http://theconversation.com/lost-in-space-australiadwindled-from-space-leader-to-also-ran-in-50-years-83310.

4 Todd Harrison, "The Future of Security in Space," in *Air Power in a Disruptive World: Proceedings of the 2018 Air Power Conference,* ed. Air Power Development Centre (Canberra: Air Power Development Centre, 2019), 44.

5 Cassandra Steer, *Australia as a Space Power: Combining Civil, Defence and Diplomatic Efforts*, National Security College Policy Options Paper Series (Canberra: Australian National University, May 2021), 2.

6 Joel Lisk and Melissa de Zwart, "Watch This Space: The Development of Commercial Space Law in Australia and New Zealand," *Federal Law Review* 47, no. 3 (2019): 450; Alice Gorman, "The Sky is Falling: How Skylab became an Australian Icon," *Journal of Australian Studies* 35, no. 4 (2011): 530.

7 Moss, "There Are Many Other Things," 443–44.

8 Brett Biddington, "Is Australia Really Lost in Space?" *Space Policy* 57 (August 2021): 5; Moss, "There Are Many Other Things," 452.

9 Duane W. Hamacher, et al., "Indigenous Use of Stellar Scintillation to Predict Weather and Seasonal Change," *The Royal Society of Victoria* 131 (2019): 24–33.

10 Hamacher, et al. "Indigenous Use of Stellar Scintillation," 25–28.

11 Trevor Leaman, "Reading the Indigenous Night Sky to Interpret Wildlife Patterns," *Wildlife Australia Magazine* 56, no. 2 (2019): 18.

12 Peter Morton, *Fire Across the Desert: Woomera and the Anglo-Australian Joint Project 1946–1980* (Canberra: Defence Science and Technology, 2017), 10–11.

13 Biddington, "Is Australia Really Lost," 4; Brett Biddington, *Skin in the Game: Realising Australia's National Interest in Space to 2025* (Giralang, ACT: Kokoda Foundation, May 2008), 10; Moss, "There Are Many Other Things," 444.

14 Moss, "There Are Many Other Things," 447–48.

15 Lisk and de Zwart, "Watch This Space," 446; Moss, "Unifying Space," 8.

16 Moss, "There Are Many Other Things," 447.

17 Morton, *Fire Across the Desert*, 135, 395; Moss, "There Are Many Other Things," 447–48.

18 Morton, *Fire Across the Desert,* 396.

19 Moss, "There Are Many Other Things," 448; Moss, "Unifying Space," 8.

20 Moss, "There Are Many Other Things," 448.

21 Biddington, "Is Australia Really Lost," 4.

22 Morton, *Fire Across the Desert*, 440–41.

23 Moss, "There Are Many Other Things," 452–53; Morton, *Fire Across the Desert*, 447–48.

24 Moss, "There Are Many Other Things," 455–57; Morton, *Fire Across the Desert*, 451–53.

25 Lisk and de Zwart, "Watch This Space," 446; Moss, "There Are Many Other Things," 457; Morton, *Fire Across the Desert*, 458.
26 "WRESAT – Weapons Research Establishment Satellite"; Morton, *Fire Across the Desert*, 483–84.
27 Morton, *Fire Across the Desert*, 473–74; Moss, "There Are Many Other Things," 457.
28 Moss, "There Are Many Other Things," 449.
29 Lisk and de Zwart, "Watch This Space," 450; Moss, "There Are Many Other Things," 449–50.
30 Moss, "There Are Many Other Things," 452.
31 Biddington, "Is Australia Really Lost," 4; Moss, "Unifying Space," 9.
32 "Canberra Deep Space Communication Complex history," NASA Website, n.d., www.cdscc.nasa.gov/Pages/cdscc_history.html.
33 Alice Gorman, "The Sky Is Falling," 529, 532, 536.
34 Tom Joyner and Isabel Moussalli, "A Space Station Crash Landed Over Esperance 40 Years Ago, Setting in Motion Unusual Events," *ABC News* (Online), 12 July 2019, www.abc.net.au/news/2019-07-12/four-decades-on-from-skylabs-descent-from-space/11249626.
35 Alice Gorman, "The Sky Is Falling," 530, 536.
36 Biddington, *Skin in the Game,* 12–13.
37 Moss, "Unifying Space," 9.
38 Lisk and de Zwart, "Watch This Space," 446.
39 Steer, "Australia as a Space Power," 1; Malcolm Davis, "Towards a Sovereign Space Capability for Australia's Defence," *The Strategist*, 3 August 2020, www.aspistrategist.org.au/towards-a-sovereign-space-capability-for-australias-defence/; Moss, "Unifying Space," 7; Department of Defence, *2020 Defence Strategic Update* (Canberra: Commonwealth of Australia, 2020), 38–39.
40 Biddington, "Is Australia Really Lost," 4.
41 Moss, "Unifying Space," 9.
42 Standing Committee on Economics, *Lost in Space? Setting a New Direction for Australia's Space Science and Industry Sector* (Canberra: Commonwealth of Australia, 2008), vii.
43 Department of Defence, *Defending Australia in the Asia Pacific Century: Force 2030* (Canberra: Commonwealth of Australia, 2009).
44 Department of Parliamentary Services, *Budget Review 2009–10* (Canberra: Commonwealth of Australia, 2009), 95–96.
45 Biddington, "Is Australia Really Lost," 2.
46 Biddington, "Is Australia Really Lost," 2–3; Steer, "Australia as a Space Power," 2.
47 Australian Space Agency, "Australian Space Agency Charter," Australian Space Agency Website, 2018, www.industry.gov.au/sites/default/files/2018-10/australian-space-agency-charter.pdf?acsf_files_redirect.
48 For a discussion of the changes included in the *Space (Launches and Returns) Act 2018* (Cth), see: Jonathon Hetherington, Spiro Kalavritinos, and Chris Case, "The Launches and Returns Act: One of the Most Significant Updates to the Space Activities Act Since its Implementation," *Allens* Website, 26 September 2019, www.allens.com.au/insights-news/insights/2019/09/the-launches-and-returns-act-one-of-the-most-significant-updates-to-the-space-activities-act-since-its-implementation/.
49 Biddington, "Is Australia Really Lost," 3; Lisk and de Zwart, "Watch This Space," 446–47, 449–52.
50 Steer, "Australia as a Space Power," 3.
51 Counter-space technologies encompass a wide range of capabilities that deny or destroy adversary space capabilities and can either be on-orbit or terrestrial in nature. These broadly fit into four categories; kinetic physical (i.e., missiles), non-kinetic physical (i.e., high-energy weapons), electronic (i.e., jamming), and cyber.
52 See: Department of Foreign Affairs and Trade, "Australia-United States Space Situational Awareness Partnership AUSMIN 2010," Department of Foreign Affairs and

Trade Website, 2010, www.dfat.gov.au/geo/united-states-of-america/ausmin/Pages/australia-united-states-space-situational-awareness-partnership; Moss, "Unifying Space," 11.

53 Peter Dutton, "The Australia-U.S. Ministerial Consultations Joint Statements: An Unbreakable Alliance for Peace and Prosperity," Department of Defence Ministers Website, 17 September 2021, www.minister.defence.gov.au/minister/peter-dutton/statements/australia-us-ministerial-consultations-joint-statement-unbreakable.

54 Moss, "Unifying Space," 11.

55 Department of Defence, *2020 Defence Strategic Update* (Canberra: Commonwealth of Australia, 2020), 38–39; Department of Defence, *Annual Report 2019–20* (Canberra: Commonwealth of Australia, 2020), 96.

56 Davis, "Towards a Sovereign Space Capability for Australia's Defence."

57 Malcolm Davis, "ADF Space Command is the Right Next Step for Australian Space Power," *The Strategist,* 5 May 2021, www.aspistrategist.org.au/adf-space-command-is-the-right-next-step-for-australian-space-power/; Charbel Kadib, "Space Division Officially Established," *Defence Connect,* 19 May 2021, www.defenceconnect.com.au/key-enablers/8069-space-division-officially-established.

58 Steer, "Australia as a Space Power," 2.

59 Darin Lovett, "ADF Space Operations: Re-focusing the Military Lens," *The Strategist,* 29 March 2018, www.aspistrategist.org.au/adf-space-operations-re-focusing-military-lens/.

60 Sarah Collins, "From Childhood Fascination to Stellar Career," *Defence News,* 17 August 2020, https://news.defence.gov.au/people/childhood-fascination-stellar-career.

61 Space Situational Awareness is the ability to survey the space environment. This involves tracking of space objects (including space debris) and their activities, monitoring space weather for potential hazardous events such as solar flares, and identifying potential threats to space-based assets or terrestrial locations. These threats encompass events such as solar flares damaging satellites, collisions of debris or other satellites, and space-object re-entry over populated areas. "First Buccaneer Satellite Mission Hailed as a Success," *Australian Defence Magazine,* 23 August 23, 2018, www.australiandefence.com.au/defence/cyber-space/first-buccaneer-satellite-mission-hailed-as-a-success.

62 Collins, "From Childhood Fascination to Stellar Career."

63 One unit or 'U' sized satellite measures 10 cm × 10 cm × 10 cm. See: Ian Cartwright, *Structural Stability Assessment of the High Frequency Antenna for Use on the Buccaneer CubeSat in Low Earth Orbit* (Edinburgh, South Australia: Defence Science and Technology Organisation, 2014), 1.

64 UNSW Canberra, "M1," UNSW Canberra Space Website, n.d., https://unsw.adfa.edu.au/unsw-canberra-space/missions/m1.

65 Space Environment Research Centre, "Research Program Four Space Segment," Space Environment Research Centre Website, n.d., www.serc.org.au/space-segment.

66 "UNSW and RAAF Launch M2 Pathfinder Satellite," *Australian Defence Magazine,* 16 June 2020, www.australiandefence.com.au/defence/cyber-space/unsw-and-raaf-launch-m2-pathfinder-satellite.

67 See, for example: Biddington, "Is Australia Really Lost," 2; Moss, "Unifying Space," 12, 14.

68 See, for example: G. le Conzannet, et al., "Space-based Earth Observation for Disaster Risk Management," *Surveys in Geophysics* 41, no. 6 (2020): 1210.

69 Linda Reynolds, "Operation Bushfire Assist Concludes," Department of Defence Ministers Website, 26 March 2020, www.minister.defence.gov.au/minister/lreynolds/media-releases/operation-bushfire-assist-concludes.

70 Christine Lunsford, "Australia's Deadly Wildfires in Photos: The View from Space," *Space* Website, 16 January 2020, www.space.com/australia-wildfires-satellite-images-2019-2020.html; G. le Conzannet, et al., "Space-based Earth Observation for Disaster Risk Management," 1210; Australian Space Agency, *Earth Observation from*

Space: Roadmap 2021–2030 (Adelaide, SA, November 2021), 8, www.industry.gov. au/sites/default/files/January%202022/document/advancing_space_earth_observa tion_from_space_roadmap_2021-20303.pdf.

71 Defence Science Technology Group, "New era in Defence Satellite Communications Research," *DST News*, 28 June 2021, www.dst.defence.gov.au/news/2021/06/28/ new-era-defence-satellite-communications-research.

72 Optus, "Optus C1 Satellite Successfully Launched," *Media Release*, 12 June 2003, www.optus.com.au/about/media-centre/media-releases/2003/06/optus-c1-satellite-successfully-launched; Optus, "Optus C1: An Australian Hotbird," Optus Website, n.d., www.optus.com.au/about/network/satellite/fleet/c1.

73 Marise Payne, "Department of Defence Extends Optus Satellite Contract," *Media Release*, 1 June 2017, www.minister.defence.gov.au/minister/marise-payne/media-releases/department-defence-extnds-optus-satellite-contract.

74 Harrison, "The Future of Security in Space," 46.

75 Defence Science and Technology Group, "Cortex Keeps an Eye on the Sky," *Media Release*, 14 May, www.dst.defence.gov.au/news/2019/05/14/cortex-keeps-eye-sky.

76 Nigel Pittaway, "Defence Rethinks Space Surveillance Roadmap," *Australian Defence Magazine*, 9 September 2021, www.australiandefence.com.au/defence/cyber-space/ defence-rethinks-space-surveillance-roadmap.

77 Ibid.

78 "New Space Facility at Exmouth Takes First Pictures," *Australian Defence Magazine*, 27 April 2020, www.australiandefence.com.au/defence/cyber-space/new-space-facility-at-exmouth-takes-first-pictures.

79 Pittaway, "Defence Rethinks Space Surveillance Roadmap."

80 "HENSOLDT Australia Launch in Tasmania," HENSOLDT Website, 21 July 2021, www.hensoldt.net/news/hensoldt-australia-launch-in-tasmania/. HENSOLDT is a German company with branches in Australia.

81 Kevin Wilcox, "Japan Aerospace Exploration Agency Recovers Asteroid Sample in Australia," *NASA Appel*, 10 December 2020, https://appel.nasa.gov/2020/12/10/ japan-aerospace-exploration-agency-recovers-asteroid-sample-in-australia/.

82 Joseph Brookes, "Australia's Biggest Rocket Launch in 40 Years Set for Friday," *Innovation Aus*, 9 September 2021, www.innovationaus.com/australias-biggest-rocket-launch-in-40-years-set-for-friday/.

83 Matthew Strong, "Australia Abandons Taiwan's TiSPACE Rocket Project after 3 Failed Launches," *Taiwan News*, 16 September 2021, www.taiwannews.com.tw/en/ news/4294197.

84 Ruth Harrison, "Black Sky Aerospace Launch Australian Developed Rocket," Space Australia Website, 24 November 2021, https://spaceaustralia.com/news/black-sky-aerospace-launch-australian-developed-rocket.

85 Steer, "Australia as a Space Power," 1; Lovett, "ADF Space Operations: Re-focusing the Military Lens."

86 Steer, "Australia as a Space Power," 2.

87 For a discussion on the Woomera Manual and the broader context of international law in the space domain, see: Cassandra Steer, "The Woomera Manual: Legitimising or Limiting Space Warfare," in *Military Space Ethics,* ed. Nikki Coleman (Havant, UK: Howgate Publishing, 2022), 179–200; Cassandra Steer, "Star Laws: The Role of International Law in Regulating Civil and Military Space Activities," in *Military Space Ethics,* ed. Nikki Coleman (Havant, UK: Howgate Publishing, 2022), 159–78.

88 "The Woomera Manual," The University of Adelaide Website, 1 July 2022, https://law. adelaide.edu.au/woomera/about.

89 Steer, "The Woomera Manual," 179–80.

90 Ibid., 180.

91   Greene, "RAAF Planning for New Military Space Command as it Celebrates 100th Anniversary"; Davis, "ADF Space Command is the Right Next Step for Australian Space Power."

92   Davis, "ADF ADF Space Command is the Right Next Step for Australian Space Power"; Steer, "Australia as a Space Power," 2.

93   Davis, "ADF ADF Space Command is the Right Next Step for Australian Space Power."

94   Brian Weeden and Victoria Samson, *Global Counterspace Capabilities: An Open Source Assessment* (Broomfield, CO: Secure World Foundation, April 2021), xxxi, https://swfound.org/media/207162/swf_global_counterspace_capabilities_2021.pdf.; Center for Strategic and International Studies, "Counterspace Weapons 101," *Aerospace Security Project,* 28 October 2019, https://aerospace.csis.org/aerospace101/counterspace-weapons-101; Rajeswari Pillai Rajagopalan, "Russian ASAT Test Highlight Urgent Need for Space Governance Negotiations," *The Diplomat,* 19 November 2021, https://thediplomat.com/2021/11/russian-asat-test-highlights-urgent-need-for-space-governance-negotiations/.

95   Rajagopalan, "Russian ASAT Test Highlight Urgent Need for Space Governance Negotiations"; Brian Weeden, "2007 Chinese Anti-Satellite Test Fact Sheet," *Fact Sheet,* 23 November 2010, https://swfound.org/media/9550/chinese_asat_fact_sheet_updated_2012.pdf; Marco Langbroek, "Why India's ASAT Test Was Reckless," *The Diplomat,* 30 April 2019, https://thediplomat.com/2019/05/why-indias-asat-test-was-reckless/; Laura Grego, "A History of Anti-Satellite Programs," *Union of Concerned Scientists,* January 2012, 12–13, www.ucsusa.org/sites/default/files/2019-09/a-history-of-ASAT-programs_lo-res.pdf.

96   Leonard David, "Space Junk Removal is Not Going Smoothly," *Scientific American,* 14 April 2021, www.scientificamerican.com/article/space-junk-removal-is-not-going-smoothly/.

97   Chloe Weiner, "New Effort to Clean Up Space Junk Reaches Orbit," *NPR,* 21 March 2021, www.npr.org/2021/03/21/979815691/new-effort-to-clean-up-space-junk-prepares-to-launch.

98   Weeden and Samson, *Global Counterspace Capabilities,* 1–22.

99   Center for Strategic and International Studies, "Counterspace Weapons 101"; Weeden and Samson, *Global Counterspace Capabilities,* 1–22.

100   Air Power Development Centre, *AFDN 1–19 Air-Space Integration* (Canberra: Royal Australian Air Force, 2019), 68.

101   Peter Dutton, "Defence Explores Options for Space Electronic Warfare," *Media Release,* 29 July 2021, www.minister.defence.gov.au/minister/peter-dutton/media-releases/defence-explores-options-space-electronic-warfare#:~:text=A%20Space%20Electronic%20Warfare%20capability,use%20of%20the%20space%20domain.

102   Malcolm Davis, "Australia and the US in Space – Ready for Lift off?" *The Strategist,* 8 July 2020, www.aspistrategist.org.au/australia-and-the-us-in-space-ready-for-lift-off/.

103   U.S. Space Command Public Affairs, "Combined Space Operations Initiative Welcomes France and Germany," US Space Command Website, 13 February 2020, www.spacecom.mil/Newsroom/News/Article-Display/Article/2083368/combined-space-operations-initiative-welcomes-france-and-germany/; Davis, "Australia and the US in Space – Ready for Lift off?"

104   Steer, "Australia as a Space Power," 3.

105   Australian Space Agency, *Earth Observation from Space: Roadmap 2021–2030.*

106   James Purtill, "Australia's First Satellite That Can Help Detect Bushfires within One Minute of Ignition set for Launch," *ABC News,* 14 March 2021, www.abc.net.au/news/science/2021-03-14/bushfires-detecting-them-from-space-fireball-satellite-launch/13203470.

References

Government sources

Air Power Development Centre. *AFDN 1–19 Air-Space Integration*. Canberra: Royal Australian Air Force, 2019.

Department of Defence. *2020 Defence Strategic Update*. Canberra: Commonwealth of Australia, 2020. www.defence.gov.au/about/publications/2020-defence-strategic-update.

Department of Defence. *Annual Report 2019–20*. Canberra: Commonwealth of Australia, 2020.

Department of Defence. *Defending Australia in the Asia Pacific Century: Force 2030*. Canberra: Commonwealth of Australia, 2009.

Department of Parliamentary Services. *Budget Review 2009–10*. Canberra: Commonwealth of Australia, 2009.

Standing Committee on Economics. *Lost in Space? Setting a New Direction for Australia's Space Science and Industry Sector*. Canberra: Commonwealth of Australia, 2008.

Published sources

Australian Space Agency. *Earth Observation from Space: Roadmap 2021–2030*. Adelaide, SA: Australian Space Agency, November 2021.

Biddington, Brett. *Skin in the Game: Realising Australia's National Interests in Space to 2025*. Giralang, ACT: Kokoda Foundation, May 2008.

———. "Is Australia Really Lost in Space?" *Space Policy* 57 (August 2021): 1–8.

Cartwright, Ian. *Structural Stability Assessment of the High Frequency Antenna for Use on the Buccaneer CubeSat in Low Earth Orbit*. Edinburgh, South Australia: Defence Science and Technology Organisation, 2014.

Gorman, Alice. "The Sky is Falling: How Skylab Became an Australian Icon." *Journal of Australian Studies* 35, no. 4 (2011): 529–46.

Grego, Laura. "A History of Anti-Satellite Programs." *Union of Concerned Scientists*, January 2012. www.ucsusa.org/sites/default/files/2019-09/a-history-of-ASAT-programs_lores.pdf.

Hamacher, Duane W., John Barsa, Segar Passi, and Alo Tapim. "Indigenous Use of Stellar Scintillation to Predict Weather and Seasonal Change." *The Royal Society of Victoria* 131 (2019): 24–33.

Harrison, Todd. "The Future of Security in Space." In *Air Power in a Disruptive World: Proceedings of the 2018 RAAF Air Power Conference*, edited by Air Power Development Centre, 43–48. Canberra: Air Power Development Centre, 2019.

le Conzannet, G., M. Kervyn, S. Russo, C. Ifejika Speranza, P. Ferrier, M. Foumelis, T. Lopez, and H. Modaressi. "Space-based Earth Observation for Disaster Risk Management." *Surveys in Geophysics* 41, no. 6 (2020): 1209–35.

Leaman, Trevor. "Reading the Indigenous Night Sky to Interpret Wildlife Patterns." *Wildlife Australia Magazine* 56, no. 2 (2019): 18–20.

Lisk, Joel, and Melissa de Zwart. "Watch this Space: The Development of Commercial Space Law in Australia and New Zealand." *Federal Law Review* 47, no. 3 (2019): 444–68.

Morton, Peter. *Fire Across the Desert: Woomera and the Anglo-Australian Joint Project 1946–1980*. Canberra: Defence Science and Technology, 2017.

Moss, Tristan. "'There Are Many Other Things More Important to us Than Space Research': The Australian Government and the Dawn of the Space Age 1956–62." *Australian Historical Studies* 51, no. 4 (2020a): 442–58.

———. *Unifying Space: Australia Needs a Whole of Government Space Policy.* Black Swan Strategy Paper Series. Crawley, Western Australia: University of Western Australia Defence and Security Institute, 2020b.

Steer, Cassandra. *Australia as a Space Power: Combining Civil, Defence and Diplomatic Efforts.* National Security College Policy Options Paper Series. Canberra: Australian National University, May 2021.

———. "Star Laws: The Role of International Law in Regulating Civil and Military Space Activities." In *Military Space Ethics*, edited by Nikki Coleman, 159–78. Havant, UK: Howgate Publishing, 2022a.

———. "The Woomera Manual: Legitimising or Limiting Space Warfare." In *Military Space Ethics*, edited by Nikki Coleman, 179–200. Havant, UK: Howgate Publishing, 2022b.

Waldrop, Elizabeth S. "Integration of Military and Civilian Space Assets: Legal and National Security Implications." *Air Force Law Review* 55 (2004): 157–232.

Weeden, Brian, and Victoria Samson. *Global Counterspace Capabilities: An Open Source Assessment.* Broomfield, CO: Secure World Foundation, April 2021.

Part Five

Future Directions

16 Can the Australian Defence Force become the most uncrewed and autonomy-enabled defence force in the world?

Keirin Joyce[1]

For more than a decade, pundits within Defence circles have been discussing and speculating on the 'tidal wave' of robotic, uncrewed, and autonomous technologies that are about to proliferate; but, just like any new, 'game-changing' technology, it has not arrived as quickly as expected. However, that is not necessarily a problem because the Australian Defence Force (ADF) is not yet ready for it. The debate about the Internet of Military Things (IoMT),[2] autonomy, and robotic/ autonomous weapons is rife among defence professionals, academics and the wider community, but solid conclusions are lacking. With these pieces of the puzzle unsolved and, in some cases, not even defined, it is beneficial that time is taken to introduce these technologies gradually. This chapter principally discusses the two primary reasons why the ADF is not ready for the proliferation of uncrewed and autonomous technology. First, the ADF does not know what to do with such technology, and second, it is not planning for the step change from automated to autonomous. This chapter argues that the crash of the tidal wave can afford to wait a little longer.

Uncrewed aerial systems

UAS, UAV, RPAS, drone—a broad array of terminology is used interchangeably by the ADF and Australian media for what the ADF generally describe as Uncrewed or Uninhabited Aerial System (UAS).[3] A UAS is all the equipment and supporting fundamental inputs to capability (FIC)[4] that enable an air vehicle (AV) (the aircraft) to undertake operations controlled by an operator on the ground, known as a Ground Control Station (GCS) (the cockpit). Broadly, there are two types of UAS: systems where the command and control link between the GCS and AV is a line of sight (LOS) radio link direct from the GCS to the AV, and systems where the command and control link is through a satellite (beyond-line-of-sight, or BLOS). This physical distinction dictates the range of the AV. LOSs are short-range, meaning the GCS must be within 200 km of the AV; however, BLOS can operate anywhere in the world, supported by the satellite system. Australia's GCS is located in Adelaide, and AV can be deployed globally. When range becomes global, the AVs generally become larger as they need to carry more fuel to increase their endurance. Consequently, LOS UASs are operated by Services

DOI: 10.4324/9781003230656-22

like the Australian Army (Army) and the Royal Australian Navy (RAN) who seek to generate UAS effects at tactical ranges. In contrast, BLOS UAS are employed by the Royal Australian Air Force (RAAF), which seeks to generate UAS effects at operational and strategic ranges.

UASs are important to the ADF as a middle-power defence force for two reasons: efficiency and risk. UASs generate efficiency and mass by placing robotic systems in the battlespace and utilising technology to gain a capability advantage across the spectrum of employment, from domestic and humanitarian emergency response through to high-end conflict. They allow Australia's smart people to operate smart equipment remotely and provide the ADF's combat forces with a force-multiplier effect, making the people on the ground, in the air and on/under the water more effective. Secondly, the risk to ADF operating personnel is reduced by removing people from within aircraft and placing them in a remote GCS. These two aspects have driven the ADF's adoption of UAS for the best part of 30 years and remain so today.

The ADF and UAS

The ADF professed the tidal wave of robotic technology proliferation in the 2004 ADF Unmanned Aerial Systems (UAS) Roadmap.[5] The UAS Roadmap utilised Joint Project 129 Phase 2 (JP129-2), which provided Tactical UAS (TUAS) in support of the Brigade or Joint Task Force Commander as the starting point for the development of UAS and their introduction into service within the ADF. It also foresaw an ultimate requirement for a fully tiered/layered capability set from Small UAS (SUAS), which would be used by an infantry section, up to strategic UAS like the Global Hawk, all of which would be delivered by 2020. When we fast forward to 2022, we can see that the UAS Roadmap has proven itself in foretelling the future, but that crash of the tidal wave in 2020 has not yet happened: the ADF is a few years behind schedule.

The ADF has not been idle, though: the tide swelled in the 2010s, with the Army, the RAAF, and the RAN testing the waters to inform themselves of the specifics of what they need. Army has been the most successful in these efforts. Due to delays in JP129-2, the Army rapidly acquired the Scaneagle TUAS and learnt lessons over 45,000 operational hours. This experience ultimately led to the RQ-7B Shadow 200 being delivered under JP129-2, directly onto operations in Afghanistan, where it quickly passed 10,000 operational flying hours. At the small end of capability, Army experimented with the Codarra, Raven, Skylark, and Wasp SUAS before deciding the RQ-12 Wasp AE would be the SUAS used to equip every combat team under Land 129 Phase 4A (L129-4A). The Army also trialled the Black Hornet Nano UAS (NUAS) before acquiring it under Army Minor Project 024.33 to equip its combat platoons.

The Army is not alone in its successful work-up; the RAAF and RAN have slowly followed. The RAAF operated the Heron medium-altitude long-endurance (MALE) UAS in support of Operation Slipper (the ADF contribution to the war in Afghanistan) land forces for the best part of a decade, accumulating tens of thousands of hours of experience throughout that time. The RAAF has also embedded personnel in the US Air Force and the Royal Air Force armed MALE UAS units. These efforts were valuable in informing Project Air 7003 (A7003) (Armed MALE UAS), which was scheduled to deliver in the mid-2020s having selected the MQ-9B SkyGuardian aircraft, the latest variant born of the MQ-1 Predator and MQ-9A Reaper family of Armed MALE UAS.[6] Even before Heron operations commenced, the RAAF collaborated with the US Navy on the MQ-4C Triton programme to ensure Australia's high-altitude long-endurance (HALE) UAS requirements would be met through Project Air 7000 Phase 1B (A7000-1B) (Maritime Intelligence, Surveillance, Reconnaissance and Response), also scheduled to deliver in the mid-2020s. The MQ-4C Triton provides persistent ISR to complement the crewed P-8A Poseidon, and these two platforms replace the AP-3C Orion Maritime Patrol and Response aircraft fleet. The Triton is designed for persistent surveillance of the ocean surface and requires other (armed) assets to prosecute targets it finds and identifies. On reflection, the RAAF's UAS efforts have been slow, lightly resourced, and measured.

In contrast to the Army and the RAAF, the RAN—arguably Australia's longest-running UAS operator thanks to its use of the Jindivik target drone from the 1950s—has been slower in engaging with contemporary UAS projects. They have had no operational imperative to do so, as their surface combatant ships already had organic, embarked aviation assets to extend their reconnaissance range: their crewed Seahawk helicopters. In contrast, the Army could not deploy the Tiger Armed Reconnaissance Helicopter to the Middle East; the capability was immature, and the RAAF lacked a persistent land ISR and strike support aircraft in inventory. The RAN has, however, hit its stride in recent years with deployed trials of the ScanEagle in a Maritime TUAS (MTUAS) configuration and the first trials of a vertical take-off and landing MTUAS capability: the S-100 Camcopter. The RAN's trial efforts will inform its Sea 129 Phase 5 (S129-5) project, which is scheduled to deliver its in-service MTUAS capability from the mid-2020s.

With this brief history and appreciation of the current situation in mind, it is clear the Army is well ahead of the other two services. The Army's relative success can be attributed to three considerations: operational demand, relatively low-cost options, and a lack of any competing aircraft acquisitions. Identifying these three areas is crucial if the ADF is to explore how it can keep pace with fast-developing technology in the future, which is ultimately the purpose of this chapter. The Army's operational demand is the leading reason for its success in UAS adoption. Australia's principal ally, the United States, invested early in TUAS technology and immediately fielded these systems (the Raven SUAS, the Shadow 200 TUAS and the RQ-1 Predator MALE UAS) operationally during the Global War on Terror (GWOT). The Iraq War (also known as the Second Gulf War) provided valuable development experience to these efforts, and GWOT technologies were, therefore, relatively mature in the 2000s. The United Kingdom similarly fielded in-service TUAS (the Phoenix TUAS) early in the GWOT. As a minor contributor to the coalition and without TUAS, the Australian Army quickly appreciated the value of these systems, which provided persistent and organic surveillance and reconnaissance. The Army also quickly realised that they could not borrow from anyone else if they did not bring their own TUAS.

The land-based TUAS employed by the Army are also relatively cheap: delivering a SUAS into use costs single-digit millions of dollars, and a TUAS costs tens of millions—cheap when weighed against the billions of dollars used to fund RAN and RAAF programmes. Furthermore, the Army had no competing (expensive) aircraft acquisitions as it moved into the UAS sphere: UAS were the new Army aeroplanes on the block in the 2010s. In contrast, for the RAAF, Heron was in-use in parallel to acquisitions of F-35A, C-17A, F/A 18F, EA-18G, KC-30A, the P-8A, and the C-27J. Likewise, the RAN's experimentation with UAS was carried out parallel to MH-60R and the Joint Helicopter School EC-135 (not to mention ships and submarines). For the RAAF and the RAN, UAS have simply not been the priority of effort in acquisition and development.

Nonetheless, a swell of effort has occurred across the services, and the sum of this swell has resulted in a Defence Integrated Investment Plan and Force Structure Plan that is scheduled to, in another five or so years, deliver one of the most uncrewed aerial defence forces in the world.[7] A telling year in this tidal chart was 2018, as the Army undertook a six-fold increase in its UAS fleet, from 150 AVs to 900. The ADF's current list of uncrewed aerial capabilities in development includes:

- Army: L125–4 – NUAS to the Combat Section;
- Army: L3025 – UAS for chemical detections;
- Army: L129-4B – Replacement SUAS to the Combat Team;
- Army: L4120 – UAS tactical Signals Intelligence effects;
- Army: JP129-2 to be replaced by L129-3 – TUAS to the Brigade/Joint Task Force;
- Army: L4503 – Armed Reconnaissance Helicopter replacement involving a mix of crewed and uncrewed platforms;
- RAN: S129-5 – MTUAS for Surface Combatants;
- RAAF: A7003 – Armed MALE UAS; and
- RAAF: A7000-1B – Maritime Patrol and Response aircraft replacement involving P-8A Poseidon and MQ-4C Triton HALE UAS.

Next steps

What does this reflection on 2004 to now mean? It can be argued that we now know when the uncrewed aerial systems 'tidal wave' will crash: the mid-2020s. This chapter outlines the tidal predictions of the early-2000s, the swelling of the tide over the past decade, the surge the ADF is currently experiencing, and the crest breaking in the mid-2020s. Considering this progression, we can see that the ADF has all but solved the challenge of getting this automated aerial technology into service, but is the ADF ready for it? The ADF arguably has only a few years left to decide how it can most effectively employ UAS (not simply acquire them most effectively) and efficiently use the vast volume of decision, intelligence, and targeting data UAS will yield.

The most immediate issue is: how can the ADF turn fifth-generation, leading-edge platforms into a world-leading warfighting capability? Within this issue are many contingent questions that must be answered. How will the many UAS capabilities interact, interoperate and interface with crewed platforms? Are whole-of-government standards in place for still imagery; motion imagery; synthetic aperture radar imagery; hyperspectral imagery; and signals intelligence collection; classification; processing, exploitation, dissemination (PED); and archiving? Are information technology networks and bearers in place to handle the volume? Can the ADF effectively undertake PED over restricted, secret, and top-secret networks or the IoMT in real time? Once decision/intelligence/targeting product is disseminated, are the tactical users able to receive and action it quickly enough for the fast pace of the modern battlefield? If not, the ADF will continue to undertake close air support in the second-generation style of the Second World War, Korean War, and Vietnam War, with UAS operators limited to talking to ground callsigns over VHF radio voice communications. Most experts would argue this would not be good enough when weighed against the sizable financial investment. Solving these most immediate issues is a large, joint endeavour that will require a championed, very well-resourced joint approach. The Defence Integrated Investment Plan allocates the work parallel to the UAS projects scheduled to deliver in the mid-2020s, but time will tell whether it can break the mould of Defence IT projects and deliver on schedule.

Over the next five years, the ADF must also ask: what is next? The 2004 ADF UAS Roadmap plotted a path for uncrewed systems that are predominantly human-in-the-loop, automated, 1:1, operator:platform systems; it has taken 20 years to be realised in the mid-2020s. This suggests the ADF needs to be writing the next roadmap for autonomous systems now (or even ten years ago). This process started with the 2020 Concept for Robotic and Autonomous Systems, but the concept does not go so far as to articulate the roadmap or implementation plan for those technologies.[8] The next roadmap should discuss the systems that will replace or spiral upgrade the current uncrewed systems; the systems that will be human-on-the-loop (or even off-the-loop), autonomous (and to what level), 1:X/ many systems. These systems will move beyond the status quo of Intelligence, Surveillance, Targeting, Acquisition, Reconnaissance and Electronic Warfare (ISTAREW) platforms. New roles will include combat service support, logistics, casualty evacuation, pseudo satellites, swarming and teaming systems, tactical combat, and loyal wingmen. More effort is needed now to articulate what this means for the ADF.

Conclusion

Recent experience indicates that the ADF will take 20 years to go from roadmap to tidal wave crash. Most technologists are predicting effective autonomy technology by 2030. If this is the case, the ADF is already running late with the 'ADF

Autonomous Systems Roadmap'. The authors of that document should be reading, researching, crafting, and drafting now. The ADF faces the risk of being left behind in an autonomy arms race. Alternatively, the ADF might execute such a roadmap faster, but this would require acquisition processes that could move at a much higher tempo than the slower pace to which Defence is accustomed. Is the Department of Defence capable of such a change? The First Principles Review of 2015, which was tasked with ensuring the Department of Defence was 'fit for purpose and able to deliver against its strategy with the minimum resources necessary', espoused this principle, directing the department towards a more streamlined acquisition cycle.[9] Still, only time will tell on that idea.

The 2020s is set to yield a fully layered UAS ISTAREW force for the ADF. The ADF was slow out of the gate regarding UAS, but that is ultimately a good thing. A few more years are needed to determine what the ADF will do with the data to turn this (expensive) uncrewed force into decision/targeting superiority and effects. The joint project tools will be available to enable success in the future. Provided they receive the prioritisation in resourcing and championship they require, these joint projects have the potential to deliver on, or close to, time. By delivering on time, the joint projects will set the conditions for the next step: taking those capabilities from automated to autonomous. How will the ADF do that? The ADF needs a plan soon, or they will find themselves playing catch-up again.

Notes

1 This chapter is based on my career in the ADF, including the past 15 years as a thought leader and change agent within the field of Uncrewed Aerial Systems within the Army and the RAAF.
2 An Internet of Things (IoT) describes an interconnected network of physical objects that carry, capture, sense, actuate, and share data with other devices (e.g., smartwatches and fitness trackers). The IoMT is a class within this broader IoT that supports combat operations and warfare. 'Things' in the IoMT can include sensors, vehicles, UAVs, and other smart technology. See for example: Greg Rowland, "The Internet of Military Things & Machine Intelligence: A Winning Edge or Security Nightmare?" *Australian Army Research Centre* website, 21 May 2017, https://researchcentre.army.gov.au/library/land-power-forum/internet-military-things-machine-intelligence-winning-edge-or-security-nightmare.
3 'Unmanned' is the term in the current ADF Glossary, however, in recent years the term 'Uncrewed' has increased in preference. Uncrewed was the term used in the most recent bout of strategic documentation including the Defence Strategic Update and Force Structure Plan of 2020. In parallel the representative industry body for UAS in Australia has recently changed its name to the Australian Association for Uncrewed Systems (AAUS).
4 FIC is defined as: Organisation, Command and Management, Personnel, Collective Training, Major Systems, Supplies, Support, Facilities and training areas, and Industry.
5 Note that 'unmanned' was the term of choice until 2019, at which time 'uncrewed' became favoured.
6 A7003 was cancelled in March 2022 in order to provide funding for Project REDSPICE, Australia's expansion of Cyber capability.
7 Department of Defence, *Integrated Investment Program* (Canberra: Commonwealth of Australia, 2016); Department of Defence, *Force Structure Plan* (Canberra: Commonwealth of Australia, 2020).

8 Australian Defence Force, *Concept for Robotics and Autonomous Systems* (Canberra: Commonwealth of Australia, 2020), 2.
9 Department of Defence, *Creating One Defence: First Principles Review* (Canberra: Commonwealth of Australia, 2015).

References

Government sources

Australian Defence Force. *Concept for Robotics and Autonomous Systems*. Canberra: Commonwealth of Australia, 2020.
Department of Defence. *Creating One Defence: First Principles Review*. Canberra: Commonwealth of Australia, 2015.
———. *Integrated Investment Program*. Canberra: Commonwealth of Australia, 2016. www.defence.gov.au/sites/default/files/2021-08/2016-Defence-Integrated-Investment-Program_0.pdf.
———. *Force Structure Plan*. Canberra: Commonwealth of Australia, 2020.

17 Space power and the vulnerabilities of satellites

James-Andre Galam[1]

Through global positioning system (GPS) capability, satellites heavily assist the Australian Defence Force (ADF) in many ways, including in positioning, targeting, command, control, and coordination. Without GPS, managing the modern battlespace domains of sea, land, air, and space becomes near impossible.[2] Protecting these assets is a critical priority, as satellites form an integral component of protecting Australia's national interests. Worldwide technological advancement has increased the potential for cyber attacks on satellites to occur. The exponential growth of artificial satellites in orbit and the development of anti-satellite missiles by other nations also threaten the already established constellation of satellites. With these factors at play, the physical and cyber security of Australian space power is already threatened and vulnerable to exploitation.[3]

The ADF requires GPS assets to protect Australian national interests efficiently, and Australian society is heavily reliant on satellites and associated capabilities such as GPS. Satellites can transmit information and communicate over long distances instantly, and functionalities such as GPS, internet services and television ensure that society operates as efficiently as possible. A threat to satellites, whether cyber or physical, presents the possibility of unprecedented chaos and catastrophe for all components of a functioning society, not only in Australia but also for other nations dependent on satellite capabilities. If an attack were to succeed, the nation would likely be severely exposed and unable to respond quickly and effectively.

Australia has recognised this vulnerability in the space domain and has begun extensive research and development into improving the nation's current space capability. With resilient and reliable space architecture, Australia would likely be better equipped to respond if its satellites were attacked, thus reducing the likelihood of the nation being rendered defenceless.

Australia's weakness in the space domain

Australia's weakness in the space domain is due to its dependency on space assets. Satellites are essential for any domain that requires modern communications, providing everything from internet access to weather forecasts and mapping services.[4] Without satellites facilitating the operation of Australian society, its economy would cease to function, as credit cards, automatic teller machines,

DOI: 10.4324/9781003230656-23

and stock market transactions rely on satellites to transmit data.⁵ To compensate for this capability gap, the Australian Government has relied on commercial and allied satellite capability to access space, highlighting a vulnerability in Australia's ability to conduct intelligence, surveillance, and reconnaissance (ISR). For example, the Optus C1 Satellite fulfils both commercial and military communication roles, and Australia has a satellite fleet that pales when compared to those of China and the United States.⁶ Historically, Australia's utilisation of space, through the acquisition and development of a satellite capability, has been limited by the budget allocated by the Australian Government.⁷ Consequently, Australia's ability to project its space power has been restricted, further contributing to the lack of capability.

Another weakness that may be exploited lies within Australian cyber security. In addition to the space domain, cyberspace permeates and connects every capability of the Australian Government. Cyber security plays an important role in protecting information and data transmitted through this domain, especially regarding satellites. During the data transmission process, this information is vulnerable to interception by adversaries, potentially leading to the leakage of sensitive information.⁸ The Australian Government's Department of Defence has recognised the rising potential threat originating from cyberspace.⁹ In 2015, the Australian Signals Directorate noted more than 1,200 cyber security breaches.¹⁰ Likewise, in the United Kingdom, authorities experienced more than 98 million cyber-attacks between 2013 and 2017.¹¹ This demonstrates that Western governments' cyber security can (and will) be breached. Australia's response to potential cyber threats has been described as 'moderate' for defence and industrial applications but 'poor' in terms of education, social management and government functions, confirming that a vulnerability exists in Australian cyber security.¹² This will have implications on the security of Australian space operations if cyber security is not developed to be accordingly resilient.

Space and satellite capabilities are critical for the ADF because it would become almost impossible for the organisation to function without space assets. Space is a dimension that provides a 'combination of perspective, persistence and freedom of operation' that cannot be achieved in the traditional dimensions of land, sea, and air.¹³ Through space, GPS allows the siphoning of information and data and enables the ADF to use Satellite Communication (SATCOM) to defend Australia's interests and assist allies in operations around the globe. GPS is a specific asset on which the ADF heavily relies, as it allows management of the battlespace through positioning, timing, coordination, command and control of platforms. SATCOM is also an important asset as it provides reliable communication between ADF elements.¹⁴ When combined, these assets allow the management of platforms across the three services of the ADF, most notably for the Royal Australian Air Force (RAAF) and Royal Australian Navy. Without GPS and SATCOM, it would become near impossible to manage these platforms and deploy them in an operational context. For example, the P-8A Poseidon maritime patrol aircraft extensively utilises SATCOM for communication. Likewise, the aviation industry relies heavily on GPS for air traffic control. Without these two assets, aviation operators must manage a greater level of risk,[15] and without satellite capability, the ADF cannot effectively pursue or defend Australia's interests, nationally or globally.

Increasingly, space has become an important focus within Australia's strategic context. This reorientation is due to the relationship between two of the world's super powers—China and the United States—and Australia's position in preserving a delicate balance between those two nations.[16] These nations, which are at the forefront of their respective space programmes combined, boast more operational satellites than any other nation.[17] Furthermore, the combination of terrorism and the rapid military modernisation of nations in Australia's immediate region validates the requirement to have Australian space power effects that can complement ISR in the neighbouring region.

The Australian Government's response

Australia has recognised the importance of space power in a security and strategic context, and efforts are now being made to create the capability to deliver this space power. The Australian Government expressed its desire to develop the nation's expertise in the space and cyberspace sectors and subsequently improve the ISR capabilities of the ADF through the delivery of space power and associated analytical imagery and targeting support.[18] The National Security Science and Technology Centre (NSSTC) has listed cyber security, intelligence and 'preparedness, protection, prevention and incident response' as scientific and technological development priorities.[19] The Australian Government has since responded with plans to invest approximately $200 billion over ten years in ADF capability, with a heavy emphasis on advancing space power effects. This investment heralded a significant transition phase for the ADF, particularly for the RAAF regarding Plan Jericho, which aims to turn the service into a modern and fully integrated combat force that can deliver air and space power effects in the information age.[20] In 2018, the Australian Space Agency commenced operations, formalising the existence of a space sector within the Australian Government,[21] while the RAAF's Air Power Development Centre relaunched as the Air and Space Power Centre in December 2020.[22]

The development of capability in space has quickly become a priority for the Australian Government as the threat to Australia's space and cyber domains grows.[23] Australia needs to be on par in terms of technological prowess with allies such as the United States, as this will greatly assist integrated and joint warfighting for current and future operations.[24] Realising how important space power is to Australia will heavily benefit the nation as it seeks to reduce the gap in space capability and overcome its vulnerability in space. The recent stand-up of Defence Space Command, the publication of the Space Power eManual, and the release of the Defence Space Strategy each demonstrate that the Australian Government and the ADF alike recognise that space is critical to the ADF's continued warfighting effectiveness.[25]

Potential threats to space assets

Analysing the potential threats to Australian and Allied space capability will provide insight into how warfare in space may look in the future. It will also demonstrate what protection methods can be utilised for space platforms to reduce harm. The space domain has not been heavily militarised, yet super powers such as China and the United States have demonstrated their capability to disable satellites via anti-satellite (ASAT) technologies.[26] However, the likelihood of these nation-states resorting to such tactics is unlikely due to the collateral damage the spread of debris can cause to all satellites.[27] If a nation like China were to engage in space warfare and neutralise satellites operated by its adversaries, the resultant debris would likely render China's fleet of satellites inoperable due to collateral damage sustained.

Nevertheless, the lack of resilient governing law and enforcement in space presents uncharted territory for international law. The United Nations' Outer Space Treaty is one of the few treaties used as the basis for space law.[28] An 'attack' could occur on a satellite, and a nation can still deny involvement. In 2009, United States satellite Iridium 33 collided with decommissioned Russian satellite Cosmos 2251. In this case, the Russian Government could deny that it intended to attack the American satellite, as it had no control of the satellite after it was decommissioned.[29] This currently grey area in international space law could conceal the intentions of potential adversary nations that purposefully decommission satellites to cause collisions with other satellites.[30] Adversaries can calculate the future positions of satellites to a high degree of precision. Therefore, this tactic could be considered an option for adversaries wishing to deny access to space capability to nations like Australia.[31] Australia and its allies will need a countermeasure to ensure enemy use of ASAT will cause as minimal damage as possible

Furthermore, the ever-present threat of space debris contributes to the highly precarious situation all satellites are experiencing. Humans will continue to launch satellites into space, which will only add to the concentration of satellites orbiting Earth and increase the likelihood of collisions. Such incidents are not without precedent. In 2009, a deactivated Russian Kosmos 2251 satellite collided with an operational American-owned Iridium 33 communications satellite. The incident caused a global stir and raised concerns that resultant debris might strike the international space station or other spacecraft.[32] At the time, the UK delegate on the Inter-Agency Space Debris Coordinating Committee and Chief Engineer at the British National Space Centre, Richard Crowther, flagged space debris as one of several 'weak links' that needed more attention.[33] As space is relatively inaccessible, debris and decommissioned satellites will continue to pose a threat to operational satellites for decades until they are either manually deorbited or their orbits decay.[34] Either option poses a physical threat to structures and humans on the ground, should debris large enough in size survive re-entry into the Earth's atmosphere. Australian satellite operators must employ a harm-reduction strategy to mitigate the effects of space debris constantly. Indeed, when launching Defence Space Command in March 2022, Chief of the Defence Force General

Angus Campbell emphasised that Australia must protect both commercial and military assets against space debris, collisions, and 'destructive acts'.[35]

In addition to protection in space, Australian satellite operators on Earth require protection from potential threats. Space platforms are not exclusively operated in space, as satellites must be operated via ground stations by a group of operators on Earth, who will command and control the operations of the satellite. The location of ground stations is usually not publicly known, but their importance presents another vulnerability that may be exploited. Adversaries could physically attack such ground stations, take over operations, obtain sensitive and critical information, and then sever communication with the satellite. Other motivations include destabilising a country via shutting down offensive and defensive systems that rely on satellites.[36] Additionally, such ground stations could be targets of cyber attacks intending to obtain sensitive information or disable satellite operations. As ground stations present adversaries with another option of attack, they must be resilient in both the physical and cyberspace domains so that the operation of satellites can continue unhindered.

If Australia's vulnerability in space is exploited, it will likely have severe consequences for national security and the safety of civilians and military personnel. Australia is heavily reliant on space capabilities, and satellites and associated operating bases are likely to be priority targets for potential adversaries. Neutralising space power effects would significantly hinder the ADF's capacity to respond, as communication and spatial awareness would be severely limited.[37] Furthermore, the ability to communicate with deployed ADF personnel would likely be near impossible, and their capacity to carry out current operations would be severely restricted. Australia's ability to conduct ISR would also be affected because these platforms rely on satellite imagery and communication networks, and the security of deployed personnel and citizens in Australia would be affected.[38] Panic would be induced among civilians because of the inability to contact and communicate. The normal day-to-day operation of society would be severely affected by the loss of basic functions such as credit cards, internet connectivity and navigation, which would freeze the nation's economy. These consequences demonstrate why Australia cannot afford to ignore its reliance on space and the subsequent capability gap.

Filling the capability gap

Australia is currently seeking to fill the space capability gap with various methods, as there is no one way to reduce Australia's vulnerability and reliance on space assets completely. With a combination of resilient and reliable solutions, Australia can be prepared for warfare in space and the resulting consequences for civilians and the military alike. These solutions fall into the 'Prepare–Prevent–Respond–Recover' methodology, a response that applies to significant threats to national security.

Improving Australia's current satellite technology is the first step to fixing the capability gap. The most viable of these technologies—the CubeSat—is being developed at the University of New South Wales (UNSW) alongside the Australian Defence Science and Technology Organisation. This microsatellite, which weighs only eight kilograms, will be able to provide the same capability as a 150-kilogram satellite.[39] A microsatellite has several physical advantages over larger satellites. Its lighter weight corresponds to a cheaper launch cost, and more microsatellite units can be added to the payload. A larger fleet also corresponds to wider coverage of Earth and additional backup satellites to handle contingencies if they occur. For coverage of the Australian landmass, a fleet of twelve 6U CubeSats covers approximately 91 per cent of all areas of interest, leading to a total payload of 96 kilograms for this fleet. A microsatellite is also less susceptible to collisions with space debris due to its smaller size, reducing the risk of damage and subsequent operational failure. CubeSats with the ability to actively remove orbital debris have also been proposed.[40] This capability represents a potential double-edged sword, as removing space debris vacates space not only for Australian and allied satellites but also for adversaries and their satellites. However, the removal of space debris will decrease the risk of collisions in space, making it a safer domain for space operations for all parties.

Australia will likely consider CubeSats as a cheap and effective way to remedy its current space capability issues due to the physical advantages CubeSats offer. Testing is underway, and a prototype—EC0—was launched by UNSW in early 2017, with the National Aeronautics and Space Administration (NASA) having already announced extensive plans to utilise CubeSats to support its Near Earth Network (NEN) CubeSat Communications.[41] However, further experimentation is required as CubeSats still need to be tested to operate in emergency conditions while also meeting civilian demand for data broadcasting.[42] The US military will likely utilise this capability, further validating why Australia should consider CubeSats, as shared technology will assist joint warfighting operations for the two allies in the future.

Australia is already preparing for the worst-case scenario where space power can no longer be generated from a military perspective. Training in environments that are less reliant on space capabilities will prepare Australia for the occurrence of this worst-case scenario. The ADF has already begun training to fight in space-denied or degraded environments, which assists the ADF in preparing for a battlespace in which Australian space capabilities are restricted. Other methods of communication and positioning, such as high-frequency radio systems and navigation and timing systems, that are not reliant on satellites are being tested in such environments.[43] Such systems would facilitate communication at an adequate level for the ADF to operate in space-denied environments, therefore preparing Australia if it were to occur.

In addition to physical protection, resilient cyber security is required to ensure command and control during the operation of satellites, as this will reduce the risk of cyber attack and theft/loss of sensitive data and information acquired and siphoned by these satellites. As mentioned above, satellites require multiple ground stations to be operated. These ground segments must be protected so the satellite can operate with maximum efficiency and protect the staff that operate them. NASA's NEN has allied and commercial ground stations located on every continent on Earth (including Antarctica), demonstrating the extensive infrastructure required to operate a fleet of CubeSats.[44] The cyber security of each ground station must be at an acceptable standard due to the increasing threat of cyber-attack.[45]

Accordingly, cyber security education and training have been deemed critical to establishing foundational knowledge in cyber security.[46] This implementation can be seen in the now compulsory Introduction to Cyber security course and the recent introduction of the Computing and Cyber security degree at the University of New South Wales, Canberra, for Australian Defence Force Academy cadets. This education is emphasised not only for the military and government but also for the rest of Australian society due to the nation's dependence on information and communications technology (ICT).[47] Education of all Australians will reduce the number of vulnerable points adversaries can exploit via cyber-attacks. In return, this will reinforce Australia's expertise in cyber security and allow the unhindered operation of Australian ICT architecture, including satellite ground stations. Australia will need to ensure that it implements these technologies as soon as they are deemed suitable to meet its space capability requirement. This will also prepare Australia's response to space warfare.

Conclusion

Having assets in the space domain is critical to Australians' daily life, from both a military and civilian perspective. It is particularly important to protect Australia's space assets and seek to develop and improve its current capability so the nation does not lose access to space and all the associated benefits that satellites provide. If Australia does fill its current capability gap, it will be able to manage the consequences of space warfare effectively should they occur. Solutions such as CubeSats and fortified cyber security will likely play a major role in maintaining the security of Australia as the likelihood of conflict in both space and cyberspace increases by facilitating the generation of space power to mitigate threats in these domains.

Furthermore, improving the potency of Australia's space assets will facilitate operations for the ADF internationally while consolidating Australia's status as a regional power in Oceania and Asia. This will assist Australia in maintaining stability within the region, especially in Southeast Asia.[48] It will also allow Australia to be a reliable force to generate and deliver this capability to its allies. Overall, the development of Australia's space capability will be crucial for reinforcing national security and protecting the nation against future threats.

Notes

1 This chapter was written while the author was an officer cadet at the Australian Defence Force Academy in 2018. It, therefore, generally reflects the state of the field at that time

and uses sources as they then stood, though some small adjustments have been made to contemporise the chapter.

2 Stephen Osborne and Andrew Jolley, "The Australian Response to Potential Space Warfare," *United Service* 67, no. 3 (Spring 2016): 21–23.

3 Department of Defence, *Defence White Paper 2016* (Canberra: Commonwealth of Australia, 2016), 51–52.

4 Jill Slay, "Training and Education for Cyber Security, Cyber Defence and Cyber Warfare," *United Service* 67, no. 3 (Spring 2016): 24–26.

5 Osborne and Jolley, "The Australian Response to Potential Space Warfare," 21–23.

6 Singtel Optus, "The Optus Satellite Fleet," *Optus* Website, accessed 20 April 2022, www.optus.com.au/about/network/satellite/fleet; Gregory Sousa, "Countries with the Most Operational Satellites in Orbit," *World Atlas* Website, 25 April 2017, www.world atlas.com/articles/countries-with-the-most-operational-satellites-in-orbit.html.

7 Gareth B. S. Neilsen, "Taking It to the Streets: Exploding Urban Myths About Australian Air Power," in *Friends in High Places: Air Power in Irregular Warfare*, ed. Sanu Kainikara (Canberra: Air Power Development Centre, 2009), 147.

8 Mick Hose, "Future ADF Satellite Communication," lecture to officer cadets at the Australian Defence Force Academy, 14 September 2018.

9 Department of Defence, *Defence White Paper 2016*, 51–52.

10 Ibid., 52.

11 Big Brother Watch, *Cyber Attacks in Local Authorities: How the Quest for Big Data Is Threatening Cyber Security* (n.p.: Big Brother Watch, February 2018), 5.

12 Slay, "Training and Education for Cyber Security, Cyber Defence and Cyber Warfare," 26.

13 Osborne and Jolley, "The Australian Response to Potential Space Warfare," 21.

14 Hose, "Future ADF Satellite Communication."

15 Royal Australian Air Force, "P-8A Poseidon," *Royal Australian Air Force* Website, n.d., www.airforce.gov.au/technology/aircraft/intelligence-surveillance-and-reconnaissance/p-8a-poseidon.

16 Department of Defence, *Defence White Paper 2016*, 15.

17 Sousa, "Countries with the Most Operational Satellites in Orbit."

18 Department of Defence, *Defence White Paper 2016*, 10.

19 Defence Science and Technology Group, *National Security Science and Technology – Policy and Priorities* (Canberra: Commonwealth of Australia, 2018), 11.

20 Royal Australian Air Force, *Plan Jericho* (Canberra: Air Power Development Centre, 2015), 3.

21 Department of Industry, Innovation and Science, "Australian Space Agency."

22 Air and Space Power Centre, "The Air and Space Power Centre Takes Flight," *Air and Space Power Centre* website, 12 December 2020, https://airpower.airforce.gov.au/news/air-and-space-power-centre-takes-flight.

23 Department of Defence, *Defence White Paper 2016*, 51–52.

24 Slay, "Training and Education for Cyber security, Cyber Defence and Cyber Warfare," 24–31.

25 Department of Defence, "Defence Soars into Space," *Department of Defence* Website, 23 March 2022, https://news.defence.gov.au/capability/defence-soars-space#:~:text=Led%20by%20Defence%20Space%20Commander,is%20the%20ultimate%20high%20ground. For the Space Power eManual, the first dedicated manual related to space power in the ADF, see: Department of Defence, *Space Power eManual: Light-Speed Edition* (Canberra: Commonwealth of Australia, 2022), www.airforce.gov.au/sites/default/files/doc/attachments/raaf-pages/213304_space_power_emanual_v1.0a.pdf. The Defence Space Strategy builds on existing strategy and planning to lay out the future of Australian Defence space efforts to 2020. See: Defence Space Command,

Australia's Defence Space Strategy (Canberra: Commonwealth of Australia, 2022), www.airforce.gov.au/our-mission/defence-space-strategy.

26 Slay, "Training and Education for Cyber Security, Cyber Defence and Cyber Warfare," 31.

27 Hose, "Future ADF Satellite Communication."

28 Osborne and Jolley, "The Australian Response to Potential Space Warfare," 22.

29 National Aeronautics and Space Administration, "Satellite Collision Leaves Significant Debris Clouds," *Orbital Debris Quarterly News* 13, no. 2 (April 2009): 1–2.

30 The ambiguity of applying international law within the context of space is a recognised issue, and one that is currently being addressed by a global team of international law experts who are collaboratively drafting "The Woomera Manual on the International Law of Military Space Activities". For more on this project, see: The University of Adelaide, *The Woomera Manual on the International Law of Military Space Activities* (Adelaide, October 2018), https://law.adelaide.edu.au/woomera/system/files/docs/ Woomera%20Manual.pdf; Cassandra Steer, "The Woomera Manual: Legitimising or Limiting Space Warfare?" ANU College of Law Research Paper No. 21.5, March 2021, https://papers.ssrn.com/sol3/papers.cfm?abstract_id=3802195.

31 Osborne and Jolley, "The Australian Response to Potential Space Warfare," 21–23.

32 Joel Achenbach, "Debris from Satellites' Collision Said to Pose Small Risk to Space Station," *The Washington Post*, 12 February 2009.

33 Quoted in Paul Marks, "Satellite Collision 'More Powerful Than China's ASAT Test," *New Scientist* website, 13 February 2009, www.newscientist.com/article/dn 16604-satellite-collision-more-powerful-than-chinas-asat-test/.

34 Osborne and Jolley, "The Australian Response to Potential Space Warfare," 22; Hose, "Future ADF Satellite Communication," 21–23.

35 Department of Defence, "Defence Soars into Space," *Department of Defence* Website, 23 March 2022, https://news.defence.gov.au/capability/defence-soars-space#:~: text=Led%20by%20Defence%20Space%20Commander,is%20the%20ultimate%20 high%20ground.

36 Defence Science and Technology Organisation, *Future Cyber Security Landscape: A Perspective on the Future* (Canberra: Commonwealth of Australia, May 2014), 15.

37 Osborne and Jolley, "The Australian Response to Potential Space Warfare," 21–23.

38 Neilsen, "Taking It to the Streets," 148.

39 Leon Stepan, Iain Cartwright, and David Lingard, "Can a Constellation of CubeSats Create a Capability? Satisfying Australia's Future Need for Multi-Spectral Imagery," *Small Satellite Conference* Website, accessed 20 April 2022, https://digitalcommons. usu.edu/smallsat/2013/all2013/24/.

40 Houman Hakima, Michael Bazzocchia, and M. Reza Emami, "A Deorbiter CubeSat for Active Orbital Debris Removal," *Advances in Space Research* 61, no. 9 (1 May 2018): 2377–92.

41 Giancarlo Bourke and Monica Lungeanu, "UNSW Launches a Cubesat into the Thermosphere!" *BlueSat: UNSW Student Space Projects* Website, 8 November 2017, http:// bluesat.com.au/unsw-launches-cubesat-stratosphere/; Scott Schaire, "Near Earth Network (NEN) CubeSat Communications" (PowerPoint presentation, NASA Small Spacecraft Community of Practice, Wallops Island, Virginia, January 2017).

42 Sara Jayousi, et al., "Flexible CubeSat-Based System for Data Broadcasting," *IEEE Aerospace and Electronic Systems Magazine* 33, no. 5–6 (May–June 2018): 56–65.

43 Osborne and Jolley, "The Australian Response to Potential Space Warfare," 23.

44 Schaire, "Near Earth Network (NEN) CubeSat Communications."

45 Department of Defence, *Defence White Paper 2016,* 51–52.

46 Slay, "Training and Education for Cyber Security, Cyber Defence and Cyber Warfare," 24.

47 Defence Science and Technology Organisation, *Future Cyber security Landscape*, 4.

48 Department of Defence, *Defence White Paper 2016*, 33.

References

Government sources

Defence Science and Technology Group. *National Security Science and Technology – Policy and Priorities*. Canberra: Commonwealth of Australia, 2018.

Defence Science and Technology Organisation. *Future Cyber Security Landscape: A Perspective on the Future*. Canberra: Commonwealth of Australia, 2014.

Defence Space Command. *Australia's Defence Space Strategy*. Canberra: Commonwealth of Australia, 2022. www.airforce.gov.au/our-mission/defence-space-strategy.

Department of Defence. *Defence White Paper – 2016*. Canberra: Commonwealth of Australia, 25 February 2016. www.defence.gov.au/whitepaper/docs/2016-defence-white-paper.pdf.

Royal Australian Air Force. *Plan Jericho*. Canberra: Air Power Development Centre, 2015. https://airpower.airforce.gov.au/sites/default/files/2021-03/AF14-Plan-Jericho.pdf.

Published sources

Big Brother Watch. *Cyber Attacks in Local Authorities: How the Quest for Big Data is Threatening Cyber Security*. N.p.: Big Brother Watch, February 2018.

Hakima, Houman, Michael C. F. Bazzocchi, and M. Reza Emami. "A Deorbiter CubeSat for Active Orbital Debris Removal." *Advances in Space Research* 61, no. 9 (May 2018): 2377–92.

Jayousi, S., S. Morosi, L. S. Ronga, E. Del Re, A. Fanfani, and L. Rossettini. 'Flexible Cubesat-Based System for Data Broadcasting.' *IEEE Aerospace and Electronic Systems Magazine* 33, no. 5–6 (May-June 2018): 56–65.

National Aeronautics and Space Administration. "Satellite Collision Leaves Significant Debris Clouds." *Orbital Debris Quarterly News* 13, no. 2 (April 2009): 1–2.

Neilsen, Gareth B. S. "Taking it to the Streets: Exploding Urban Myths About Australian Air Power." In *Friends in High Places: Air Power in Irregular Warfare*, edited by Sanu Kainikara, 117–86. Canberra: Air Power Development Centre, 2009.

Osborne, Stephen, and Andrew Jolley. "The Australian Response to Potential Space Warfare." *United Service* 67, no. 3 (Spring 2016): 21–23.

Slay, Jill. "Training and Education for Cyber Security, Cyber Defence and Cyber Warfare." *United Service* 67, no. 3 (Spring 2016).

Steer, Cassandra. "The Woomera Manual: Legitimising or Limiting Space Warfare?" ANU College of Law Research Paper No. 21. 5, March 2021. https://papers.ssrn.com/sol3/papers.cfm?abstract_id=3802195.

The University of Adelaide. *The Woomera Manual on the International Law of Military Space Activities*. Adelaide, October 2018. https://law.adelaide.edu.au/woomera/system/files/docs/Woomera%20Manual.pdf.

18 Considering the effects of disruptive technologies on air and space power[1]

Christopher Wooding

Disruption has become the buzzword of the modern military in recent years as geostrategic and social change challenge conventional thinking. Dramatic techno-logical advancements and societal changes have underscored the need to reorientate.[2] Among these advancements, cyber, space, uncrewed aerial systems (UAS), and artificial intelligence (AI) will be increasingly critical to reshaping the future character of war. Effectively employing these technologies will, in turn, demand new ideas to drive future air and space power in an uncertain future operating environment. This chapter will examine aspects of these technologies and their emerging impact on air and space power. It will consider each technology before discussing some implications for the Royal Australian Air Force (RAAF).

Since its practical implementation in the early 20th century, air power has repeatedly demonstrated its value to contemporary warfare. Similarly, space power has grown in importance to enable modern warfighting capabilities. The 2020 Defence Strategic Update highlighted the importance of air and space power to Australian national security, especially noting the shifting geostrategic environment.[3] The RAAF's plan for transformation in response to this shift has been underpinned by two strategies: Plan Jericho, a strategy released in 2015 that sought to capitalise on future high technology systems; and the more recent *Air Force Strategy*, released in 2020, which addresses the strategic directions associated with integrating air and space power within the joint force.[4] These strategies include introducing leading-edge platforms such as the F-35A Lightning II, E-7A Wedgetail, E/A-18G Growler, and the MQ-4C Triton. These strategic documents also articulate the Australian Defence Force's (ADF) intent to improve access to space systems, including developing sovereign capabilities. Although technologi-cal in nature, these acquisitions will entail the need for new concepts to enhance the contributions of air and space power to the ADF's Joint Force.

The era in which possession of advanced systems provided various advantages when facing less technologically advanced opponents in uncontested skies is quickly coming to an end. The rise of China and Russia challenges the air suprem-acy that Western nations have taken for granted in recent years. This develop-ment is reflected in new and emerging weapon systems, specifically anti-access/ area denial capabilities that employ advanced beyond-visual-range and hyper-sonic weapons, requiring greater reliance on sensors and information systems to

DOI: 10.4324/9781003230656-24

identify, track, and engage targets at increased ranges. Crucially, as information dependence increases within the battlespace, the air and space environments in the 2020s and beyond will be shaped by the cyber capabilities to enable networked and operational effects.

Cyber

Cyber capabilities continue to drive significant changes in the character of war. Albeit an umbrella term for a broad range of technologies, cyber's characteristics alter how nations conceptualise coercion and force to achieve national objectives. First, the pervasiveness of cyber capabilities means that everywhere is the front-line; there is no sanctuary from the effects of cyber warfare. Second, cyber warfare is both scalable and flexible, making it particularly dangerous given the plethora of ways it can be utilised. This scalability allows cyber to generate effects from the tactical to the grand strategic level, while flexibility generates asymmetric effects across the spectrum from competition to conflict. Therefore, military forces need to be prepared at all levels to defend against and utilise cyber capabilities. Third, cyber capabilities can be employed as a stand-alone strike capability, but they may also enable other strike capabilities and shape the battlespace.

Cyber warfare can be considered to deliver non-kinetic first-order effects in general, however, evidence exists of cyber capabilities achieving second- and third-order kinetic effects. Whereas kinetic actions employ physical means such as bombs, bullets, and other munitions to achieve an effect (e.g., the destruction of a weapons facility), non-kinetic actions, including cyber warfare, are generally indirect.[5] Described by Lehto and Henselmann as a 'war in bits and bytes', cyber capabilities may be used both defensively and offensively to protect against enemy cyber-attacks and keep assets operable or to 'paralyze' or destroy an enemy's ICT-based systems.[6] Non-kinetic effects enable the disruption and disablement of adversary decision-making systems through jamming sensors, disrupting information systems, or spreading disinformation, allowing reversible effects. Yet such capabilities can also have kinetic effects, including disabling key infrastructure or destroying equipment, as demonstrated when the 'Stuxnet' worm was used against Iranian nuclear processing centrifuges. In this instance, the malware virus infected the plant's computer systems, opening a backdoor into the previously offline network and allowing those behind the attack to control the plant's centrifuges. The virus allowed the plant's centrifuges to spin at a much higher rate, causing them to tear apart.[7] Together, these effects—particularly non-kinetic effects—double as critical enabling effects for the joint force, making cyber a central part of so-called grey zone operations that exploit cyber's characteristics to generate attributable and non-attributable effects below the threshold of open conflict.

Grey-zone operations often involve an informational component that employs disinformation and confusion and typically exploits cyber means, especially social media. These types of operations are likely to become an increasingly common feature of future conflict. In an air power context, exploiting cyber capabilities may become the tip of the spear. For instance, capabilities such as the E/A-18G enable electronic attacks against an adversary's defences to support the delivery of kinetic effects. Alternatively, traditional air power roles such as strike could also support the cyber campaign as the main effort. Such an idea is highlighted in the recent *Air Force Strategy*, which states:

> Creativity is required to maximise Air Force's value to the joint force in an environment of strategic competition . . . novelty and creativity are central characteristics when engaging in an environment of strategic competition, and [air and space power practitioners must] be able to apply air and space power in ways that cannot easily be anticipated by competitors or adversaries.[8]

While cyber can be used in an offensive context, it is also critical to protect against threats to maintain the integrity and security of data to enable unimpeded force projection. These various opportunities make cyber a key feature of future warfare, and militaries must excel in cyber capabilities to defend against and prosecute cyber-attacks. From a strategic perspective, Australia must prioritise the development of an indigenous cyber capability.[9] This capability needs to be underpinned by robust and comprehensive joint doctrine and concepts of operation and a specialist workforce. The RAAF's creation of a Cyber Warfare category in 2019 is an important step in this direction.[10] Without this foundation, the ability to exploit the cyber advantage will be severely curtailed. In an operational sense, the ability to project air power, in particular, will be reliant on first winning the cyber battle. As such, air power needs to adapt not only to cyber but also to complementary disruptive technologies, such as space, to maintain its utility.

Space

Space is often called the 'ultimate high ground', referring to its unique ability to physically, rather than digitally, access any part of the globe. This unique access is exploited for remote sensing, precise positioning, and communication activities. The growing dependency upon and associated vulnerability of space systems in modern warfare has increased the focus on national space capabilities and structures. In this respect, the growing operational importance of the space domain was highlighted in the 2018 announcement of the establishment of a United States Space Force.[11] For Australia, this importance has been underscored by the establishment of the Australian Space Agency in 2018 and a Defence Space Command in 2021. These entities recognise the need to consolidate national capabilities and expertise, especially in acquiring sovereign space assets. This trend reflects the growing importance of the space domain in the Australian context and its importance in facilitating growing alliance obligations and collaboration.

The risk posed by the weaponisation of space increases as adversaries compete for strategic advantage. In particular, China, Russia, and the United States have sought to dominate the space domain while developing the ability to deny an adversary a space advantage.[12] This is demonstrated by the increased focus on the counter-space role and associated capabilities.[13] In terms of denial, satellite systems are particularly vulnerable due to their predictable paths making anti-satellite (ASAT) capabilities increasingly relevant. China, Russia, and the United States have pursued the development of ASAT capabilities[14] to deny opponents access to space-based effects, which places adversaries at a major disadvantage, especially in a near-peer conflict. Weaponisation will likely emerge as terrestrial-based and satellite-to-satellite targeting capabilities. Still, these are likely to be non-destructive to prevent exponential debris growth that could lead to Kessler syndrome,[15] which would be disadvantageous to allies and adversaries alike.

Another significant trend is the increased accessibility of space, including space tourism.[16] Commercial service providers such as SpaceX and Blue Origin have demonstrated revolutionary reusable launch vehicle and crew capsule technology, significantly lowering launch costs for satellites.[17] Combining technologies that enable smaller and lighter satellites (such as CubeSats) further reduces the financial costs for nations and non-state actors to gain space access and its advantages.[18] Militarily, these advances critically support the ability for rapid reconstitution of space capability in support of operations.

Space plays a key role in connecting and enabling cyber systems and products. In this respect, Australia is heavily reliant on space systems to maintain a connection to and an awareness of the rest of the world. Without this, Australia will be less able to support regional and international security. Australia must develop some degree of sovereign space capability, which is becoming an increasingly feasible and affordable option, to avoid becoming isolated. While international partners will continue to be a key aspect of Australian policy and capability, decreasing Australian reliance on those partners will help reduce vulnerability to disruption and increase national resilience—both of which have been highlighted by the impact of the ongoing COVID-19 pandemic on global activities. A sovereign space capability would also improve the Australian military value to allies through self-reliance and interoperability.

In an air power context, an assured space advantage facilitates operational capabilities and the delivery of precision effects to support the fighting force, as reflected by the role of remote sensing and positioning, navigation, and timing (PNT) in target acquisition and the use of precision-guided munitions. This has been a growing feature of modern warfare since the first Gulf War.[19] In the future, space will be increasingly important in supporting global warfighting systems, especially for autonomous air systems such as the Boeing Loyal Wingman.[2021]

Uninhabited aerial systems

Another key development in modern conflict is the proliferation of UAS. UAS refer to autonomous rather than remote piloted systems, an important distinction for policy-makers and commanders in the field. These may include entirely uncrewed or optionally crewed platforms, allowing them to be operated as autonomous systems or standard crewed platforms. UAS offer two major advantages over traditional aircraft: range and endurance. Unlike crewed aircraft, UAS are not limited by any need to maintain a life support system. They can stay on station significantly longer than their inhabited counterparts and are limited by technical considerations, such as fuel, rather than the crew. In a strike context, this advantage allows militaries to strike further and facilitates better responsiveness to missions of opportunity than traditional aircraft. Similarly, from an air combat perspective, aircraft may be more agile given that human limitations are no longer a factor when performing manoeuvres. Without the risk to human life associated with inhabited aircraft, UAS are also expendable, if necessary.

Modern UAS vary greatly in size, complexity, and cost. Like space systems, UAS capabilities are becoming more affordable and accessible through low technology and small commercially available drones. The proliferation of commercial drones has enabled a wider variety of applications and customers, including terrorist organisations, with unforeseen effects. For instance, the 2020 conflict between Azerbaijan and Armenia in Nagorno-Karabakh demonstrated the efficacy and lethality of small drone systems.[22] Likewise, the expensive Skyguard/Patriot air defence systems failed to defend against relatively cheap drones when a drone strike was launched on a Saudi oil refinery.[23]

Australia has already begun capitalising on the scalability of drones, purchasing various types, from MQ-4Cs to Black Hornets.[24] These acquisitions are indicative of the various uses of drone technology; however, the somewhat ad hoc procurement process highlights a tendency to consider UAS as *augmenting* a large air force operating aircraft instead of considering the whole air power spectrum that includes the small unit operation of miniaturised drones up to the largest air mobility aircraft. For Australia, this would mean having a much larger set of options from which to choose in terms of acquisition and employment— both internal and external to the ADF, such as in emergency services and policing elements—which, consistent with Air Force strategy, may better serve government's requirements across the competition to conflict spectrum, rather than simply future high-end warfighting concepts.

For air power, a key UAS benefit is enhanced persistence. Operationally, persistence enables greater time over the battlespace for either surveillance or engagement, especially if this involves time-sensitive targets and targets of opportunity. The strategic benefit of this lies in addressing one of the long-standing deficiencies of air power, namely, sustaining enduring on-task responsiveness. This responsiveness could only be achieved in the past through large numbers of aircraft and crews constantly cycling to meet task requirements, limiting overall capability by negatively impacting crews and aircraft fleets. UAS allow militaries to overcome this limitation through greater flexibility, less risk to force or the mission through aircrew fatigue, and reduced fleet size and manning requirements. However, to achieve the level of autonomy needed to make UAS truly effective, they will need AI systems.

Artificial intelligence

While general AI systems have been theorised for many decades, other forms of AI are already a fundamental part of the modern world, underlying many of the systems used daily. When discussing AI, two key variables determine their influence: degree of autonomy and degree of agency.[25] Autonomy refers to the ability to do work without input, that is, to carry out a specific task without human assistance. In contrast, agency refers to the ability to act without external guidance or, more simply, the ability to choose what tasks to do. This distinction is important as it affects the legal, ethical, and trust issues that revolve around AI systems and their usage, prompting two key questions. First, is AI a driver or a supporter; that is, should it be directly in the fight, or does it merely support the war-fighter? Second, does AI require a human in the loop to operate it; if so, where?

Discussions about AI systems have generally focused on human-AI teaming to bolster the human ability with machine processing.[26] With a low degree of autonomy and agency, having a human in the loop to direct and interact with AI will significantly increase the speed of decision-making and improve decision superiority, allowing enhanced operational responsiveness on the part of human command nodes without shifting authority to machine systems. At the other end of the spectrum, AI with high autonomy and agency will be capable of directing platforms, units, or even entire operations with minimal human intervention.

In terms of air power, AI systems will likely have a high degree of autonomy to enable control of air assets, such as drones, to complete missions. New developments, such as the USAF's ARTUμ,[27] show the potential of AI to enhance combat operations of aircraft through human-machine teaming. They can optimise aircraft systems to allow the aircrew to focus on tactics,[28] which will be crucial to single-seater aircraft, but it does not make non-pilot aircrew obsolete. The introduction of AI into the cockpit may shift non-pilot aircrew to focus on coordinating drones and other assets forward of other control nodes, especially in degraded electromagnetic and non-permissive air environments, leaving the pilot or AI to focus on the aircraft itself.[29]

Despite the benefits of AI, there will always be the risk that the consequences of a rogue system are realised. For the foreseeable future, AI systems will likely still rely on some measure of human direction to determine which missions to complete and monitor mission conduct. In this sense, AI will probably have a greater initial effect beyond operations encompassing command and control, logistics, and management functions. When it comes to creating operational concepts and doctrines that integrate AI systems, it is important to understand that AI is not an end-state technology; it will continue to evolve and change. However, AI is a major enabler of other technologies, thereby underpinning cyber, space, and UAS, significantly impacting tomorrow's RAAF.

Conclusion: implications for the RAAF

This chapter has provided an overview of current technological advancements and their emerging impact on air and space power. It has considered the impact of these technologies on the future operating environment and in the context of the RAAF. These four technologies are all maturing and developing concurrently, drastically increasing the complexity of adapting to the changing air and space environment. Disruptive technologies will radically alter the air and space environments of the future. For the RAAF, this means a significant change in the way Australian air and space power operates, requiring new and innovative solutions in the transition to the next iteration of air forces and beyond. To become the future air force the RAAF aspires to be, it will be necessary to develop innovative concepts and comprehensive joint doctrine to fully realise the collective power behind cyber, space, UAS, and AI. These developments will ensure these technologies enhance the flexibility and utility inherent in air and space power and enable greater persistence, responsiveness, precision, and decision superiority.

These changes will be underpinned by the need to evolve the culture and the people within the RAAF to best integrate with emerging technologies. The effects air and space power will be required to achieve in the future to support national security must be considered. From doctrinal and workforce perspectives, how those effects are achieved also requires re-evaluation. Updating the human management component of future air forces will be critical to maintaining relevance and enabling organisational transformation to adapt to the future operating environment. In short, integrating disruptive technologies will require new thought to sustain the effective application of air and space power.

It will remain necessary to overcome various challenges, from ethical and legal to technical and intellectual, in exploiting these technologies for the future of war.[30] Overcoming these challenges will demand commitment and investment to re-envision air and space power and lead the organisation through the necessary conceptual and cultural transformations. To borrow Army parlance, staying 'future ready' demands early research, development, and investment in these critical technologies to innovate and adapt them to Australian needs and create new concepts for their employment.[31] While technology cannot solve every problem, Australia must adapt and exploit a disruptive environment; it cannot afford to be left behind as the character of war evolves.

Notes

1 This was originally written in 2018 while the author was an Officer Cadet at the Australian Defence Force Academy. It is based on a presentation for the 2018 Sir James Rowland Air Power Seminar.
2 National Intelligence Council, *Global Trends 2040: A More Contested World* (Washington, DC: National Intelligence Council, March 2021).
3 Department of Defence, *2020 Defence Strategic Update* (Canberra: Commonwealth of Australia, 2020).
4 Royal Australian Air Force, *Plan Jericho* (Canberra: Air Power Development Centre, 2015).
5 US Air Force, *Air Force Doctrine Document 2* (Washington, DC: US Air Force, 2007).
6 Martti Lehto and Gerhard Henselmann, "Non-Kinetic Warfare: The New Game Changer in the Battle Space," *Proceedings of the 15th International Conference on Cyber Warfare and Security* (March 2020): 320.
7 Michael Holloway, "Stuxnet Worm Attack on Iranian Nuclear Facilities," *Stanford University* Website, 16 July 2015, http://large.stanford.edu/courses/2015/ph241/holloway1/.

8 Royal Australian Air Force, *Air Force Strategy: Key Highlights* (Canberra: Commonwealth of Australia, 2020), 8.

9 Tim Scully, "Cyber Security and the 2016 Defence White Paper," *Security Challenges* 12, no. 1 (2016): 115–26.

10 Bel Scott, "RAAF Launches New Cyber Force," *Department of Defence* Website, 1 November 2019, https://news.defence.gov.au/technology/raaf-launches-new-cyber-force.

11 "Donald Trump Sets Goal to Create US Military Space Force by 2020," *ABC News*, 10 August 2018, www.abc.net.au/news/2018-08-10/trump-sets-goal-to-create-us-military-space-force-by-2020/10103876.

12 Loren Thompson, "Trump's 'Space Force' Motivated by Russian, Chinese Threats to Critical U.S. Orbital Systems," *Forbes*, 18 June 2018, www.forbes.com/sites/lorenthompson/2018/06/18/trump-embraces-space-force-as-russia-china-threaten-critical-orbital-systems-of-u-s/?sh=58f646b2eece.

13 Aaron Mehta, "America's Adversaries Keep Investing in Weapons to Take Out Satellites," *DefenseNews*, 30 March 2020, www.defensenews.com/battlefield-tech/space/2020/03/29/countries-keep-investing-in-weapons-to-take-out-satellites/.

14 Sandra Erwin. "U.S. Intelligence: Russia and China Will Have 'Operational' Anti-Satellite Weapons in a Few Years," *SpaceNews*, 14 February 2018, https://spacenews.com/u-s-intelligence-russia-and-china-will-have-operational-anti-satellite-weapons-in-a-few-years/.

15 Kessler syndrome describes a situation in which low-Earth orbit has become so dense with space debris that it is impossible to penetrate the debris layer and access space. It is generally seen to be caused by a rapid cascade of space debris impacting satellites and causing more debris.

16 Emily A. Margolis, "Space Tourism: Then and Now," *Smithsonian – National Air and Space Museum* Website, 25 October 2021, https://spacenews.com/u-s-intelligence-russia-and-china-will-have-operational-anti-satellite-weapons-in-a-few-years/.

17 David Z. Morris, "Is SpaceX Undercutting the Competition Even More Than Anyone Thought?" *Fortune* Website, 18 June 2017, http://fortune.com/2017/06/17/spacex-launch-cost-competition/.

18 Elizabeth Howell, "Cubesats: Tiny Payloads, Huge Benefits for Space Research," *Space.com* Website, 19 June 2018, www.space.com/34324-cubesats.html.

19 Larry Greenemeier, "GPS and the World's First 'Space War'," *Scientific American* Website, 8 February 2016, www.scientificamerican.com/article/gps-and-the-world-s-first-space-war/.

20 Loyal Wingman is an autonomous combat drone being designed by Boeing in concert with the RAAF for human-machine teaming with fighter aircraft, such as the F-35 or F/A-18F.

21 "Boeing Airpower Teaming System," *Boeing* Website, accessed 14 December 2021, www.boeing.com/defense/airpower-teaming-system/index.page.

22 Robyn Dixson, "Azerbaijan's Drones Owned the Battlefield in Nagorno-Karabakh – and Showed Future of Warfare," *The Washington Post*, 11 November 2020, www.washingtonpost.com/world/europe/nagorno-karabkah-drones-azerbaijan-aremenia/2020/11/11/441bcbd2-193d-11eb-8bda-814ca56e138b_story.html.

23 B. Hubbard, B. Karasz, and R. P. Stanley, "Two Major Saudi Oil Installations Hit by Drone Strike, and U.S. Blames Iran," *The New York Times*, 14 September 2019.

24 Baz Bardoe, "Drones and Defence: The ADF and Unmanned Aerial Systems," *Australian Aviation* Website, 7 December 2019, https://australianaviation.com.au/2019/12/drones-and-defence-the-adf-and-unmanned-aerial-systems/.

25 Johnny DiBlasi, et al., "Agency and Autonomy: Intersections of Artificial Intelligence and Creative Practice" (Unpublished Manuscript, 2020), https://dora.dmu.ac.uk/bitstream/handle/2086/20430/Agency%20and%20Autonomy.pdf?sequence=1&isAllowed=y.

26 Heather Roff, "Human Machine Teaming: Challenges, Opportunities and Gaps?" (Lecture, 2021 Air and Space Power Seminar Series, 11 June 2021), https://airpower. airforce.gov.au/videos/human-machine-teaming.
27 ARTUµ is an artificial intelligence program developed by the US Air Force and the US's Defence Advanced Research Projects Agency to support human aircrew through control aircraft systems. It is based on the µZero programme that's designed for mastering various physical and video games. ARTUµ is a modern example of human-machine teaming in combat aircraft.
28 Malcolm Davis, "The Artificial Intelligence "Backseater" in Future Air Combat," *The Mandarin*, 5 February 2021, www.themandarin.com.au/148861-the-artificial-intelli gence-backseater-in-future-air-combat/.
29 Graham Scarbro, "Naval Flight Officers' Unmanned Future," *United States Naval Institute* Website, accessed 30 October 2021, www.usni.org/magazines/proceedings/ 2021/september/naval-flight-officers-unmanned-future.
30 Linda Johansson, "Ethical Aspects of Military Maritime and Aerial Autonomous Systems," *Journal of Military Ethics* 17, no. 2–3 (2018): 140–55; Nikki Coleman, "Ethical Issues Raised by Military Uses of Space" (Lecture, 5th Sir James Rowland Seminar, University of New South Wales – Canberra, Canberra, 2019), www.youtube.com/ watch?v=pReDU9I_Uyc.
31 Rick Burr, *Army in Motion: Command Statement* (n.p.: Australian Army, October 2020), 1–2.

References

Government sources

Department of Defence. *2020 Defence Strategic Update*. Canberra: Commonwealth of Australia, 2020. www.defence.gov.au/about/publications/2020-defence-strategic-update.
National Intelligence Council. *Global Trends 2040: A More Contested World*. Washington, DC: National Intelligence Council, March 2021. www.dni.gov/files/ODNI/documents/ assessments/GlobalTrends_2040.pdf?fbclid=IwAR3wZMztcbLulYCNtEEC3cm3Xv4 lHT-fMEjuvEfABcDaL0MJZZ24i59JE1I.
Royal Australian Air Force. *Plan Jericho*. Canberra: Air Power Development Centre, 2015. https://airpower.airforce.gov.au/sites/default/files/2021-03/AF14-Plan-Jericho.pdf.
———. *Air Force Strategy: Key Highlights*. Canberra: Commonwealth of Australia, 2020. www.airforce.gov.au/sites/default/files/air_force_strategy.pdf.
US Air Force. *Air Force Doctrine Document 2*. Washington, DC: US Air Force, 2007.

Published sources

Burr, Rick. *Army in Motion: Command Statement*. N.p.: Australian Army, October 2020. www.army.gov.au/sites/default/files/2020-11/2020%20Command%20Statement_2.pdf.
Coleman, Nikki. "Ethical Issues Raised by Military Uses of Space." Lecture presented at the 5th Sir James Rowland Seminar, University of New South Wales – Canberra, Canberra, 2019. www.youtube.com/watch?v=pReDU9I_Uyc.
DiBlasi, Johnny, Carlos Castellanos, Eunsu Kang, Fabrizio Poltronieri, and Leigh Smith. "Agency & Autonomy: Intersections of Artificial Intelligence and Creative Practice." Unpublished manuscript, uploaded 12 November 2020. PDF. https:// dora.dmu.ac.uk/bitstream/handle/2086/20430/Agency%20and%20Autonomy.pdf? sequence=1&isAllowed=y.

Johansson, Linda. "Ethical Aspects of Military Maritime and Aerial Autonomous Systems." *Journal of Military Ethics* 17, no. 2–3 (2018): 140–55.

Lehto, Martti, and Gerhard Henselmann. "Non-kinetic Warfare – The New Game Changer in the Battle Space." *Proceedings of the 15th International Conference on Cyber Warfare and Security* (March 2020): 316–25.

Roff, Heather. "Human Machine Teaming: Challenges, Opportunities and Gaps?" Lecture presented at the 2021 Air and Space Power Seminar Series, online, 22 June 2021. https://airpower.airforce.gov.au/videos/human-machine-teaming.

Scully, Tim. "Cyber Security and the 2016 Defence White Paper." *Security Challenges* 12, no. 1 (2016): 115–26.

Conclusion

Jarrod Pendlebury

The preceding chapters reflect the astonishing contribution of Australians and their Indigenous forebears in developing new ways of observing, exploring, and harnessing air and space to ensure safety, security, and prosperity in our region. At its heart, this is a story of ingenuity and resilience in the face of domains that, for many thousands of years, seemed impervious to human exploitation. In many ways, the harsh, unforgiving nature of air and space reflects the Australian continent itself, so it is perhaps not surprising that Australia was among the first nations to realise the military benefit of controlling the ultimate high ground.

The first section of this book highlighted some of the lesser-known historical chapters informing the Australian approach to air and space power. In shining a light on the convergence of powered flight and the First World War, Michael Molkentin presents a theme that resonates throughout this book: the role of war – or the threat of it – as a powerful force for innovation and technical development. Later in the section, Kristen Alexander performs a neat inversion of the usual historical treatment of aviators. Her analysis of the contribution of RAAF personnel while captive as prisoners in the Second World War – through a concerted campaign of 'active disruption' – demonstrates how these efforts contributed to the broader allied air power effort. In a more abstract sense, the early chapters of the volume reinforce the indelible link between the social – positing war as the ultimate battle for the primacy of a particular social identity – and the technological. By focusing on the human side of air and space power history, Molkentin, Alexander, Yule, and Townsend give weight to the argument that the development of technology – and the organisational response it elicits – is very much a social construction. Rather than humans being shaped by technology, the social (and as Thomas Kuhn points out, political) environments in which air and space power innovation occur ultimately shape the conditions that herald new paradigms.[1] For example, in their analysis of Sir James Rowland's contribution to Australian air power, Peter Yule and Nicole Townsend demonstrate the power of the individual in shaping the strategic direction of technology. In closing his chapter, Peter Hunter brings the argument for the social construction of air and space power into sharp relief. By demonstrating the challenges that faced the RAAF during the Malayan emergency, Hunter articulates the strategic risk facing senior commanders who fail to appreciate the interdependence of political, social, and geostrategic

DOI: 10.4324/9781003230656-25

factors and their influence on military operations. Hunter's chapter offers fresh relevance in the wake of the West's ignominious withdrawal from Afghanistan, the most recent instance in which technological overmatch has proven insufficient to prevail against a determined adversary. Indeed, Russia's early difficulties following their invasion of Ukraine in February 2022 suggest this is not a trap unique to the Western military mind.

Part Two of the volume pulled at the thread of identity as it pertains to air and space power practitioners, with Jason Begley and Travis Hallen asking a question that lingers to this day: why do we need a separate air force? Interestingly, a similar debate is now taking place, leading to the emergence of 'space forces', such as the US Space Force, formed in 2019, and Australia's own Defence Space Command, formally established in March 2022. Australia is something of an anomaly among global air forces, having forged its autonomy early enough to stake a credible claim to the status of the second oldest independent air force.[2] By comparison, the US Air Force is relatively youthful, having separated from the US Army in July 1947.[3] While these outward-facing identities tend to attract broader interest, inward-focused *self*-identities play a critical role in establishing the norms, behaviours and traditions of a military organisation.[4] Self-identity, broadly defined as how aviators view themselves, comprises several individual markers that combine to sketch what, in Chapter 6, I term 'ideal characteristics'. In developing this argument, I look to basic military training as the 'theatre' in which these ideal traits are presented by senior members, prompting an obligation on the part of the junior cadets to replicate through performance. The more exclusive the content of these characteristics – highly developed upper body strength, for instance – the less potential there is for cultivating an inclusive culture in which the broadest cross-section of society is welcome. The power of self-identity in driving culture is a theme that also permeates Tom Frame's chapter, in which he analyses the ingrained preference for technical excellence in the RAAF. This pursuit of ever-greater capability has traditionally come at the expense of broader education, which focuses more on expanding *capacity*. Recent strategic messaging from the RAAF's Air Force Headquarters suggests Frame's arguments have been absorbed:

> Air Force's culture and reward system [must be] one that values leadership and strategic thinking while assuring tactical and technical excellence. The Air Force SLT [Senior Leadership Team] contains a broad mix of the best leaders and thinkers in the organisation, irrespective of their tactical and technical background.[5]

Part Three took a closer look at the culture of technology and instrumental excellence Frame argues is central to the RAAF's ethos, with each author using the intersection of technology and culture as a prism through which to analyse key moments in the RAAF's history. The chapter unfolded through a fascinating debate between authors who stand on both sides of the earlier question of whether social forces drive technological changes or technology is, by definition, socially constructed. Charles Vandepeer, Peter Layton, and Michael Spencer all present chapters informed by a fundamental view of technology as a driver for social, operational, and strategic change. Whether it be the dramatic developments in air power technology in the Second World War or the development of uninhabited aerial systems and emerging hypersonic technologies, Vandepeer, Layton, and Spencer argue from a traditional air force viewpoint that technology drives air force culture and operational perspective. Matt Hegarty presents an interesting counterpoint by arguing that the culture of air forces is often shaped by the need to address technological development, but the culture itself conditions any such response. In short, Hegarty's view is that although strong military cultures *can* follow trends sympathetic to the trajectory of technological development, they are also capable of pushing back against innovation and change.

The book shifted gear in Part Four with Deane-Peter Baker's reframing of the analysis to consider recent developments in air and space power through the lens of the human interaction of operational design. In his chapter, he points to an unsettling tendency of the 'lure of strike' supplanting the place of ethics in complex military campaigns that lack a discernible 'front line', or even a clear concept of an 'enemy'. Despite the experience gained over nearly 20 years of involvement in operations in Afghanistan, there seems little evidence that Western military thinking has evolved or expanded to consider theoretical frameworks outside the established strategic *canon*. This thread is continued by Jo Brick, who observes the West's enduring obsession with kinetic action through 'the allure of battle'. In calling for a new concept of 'manoeuvre', Brick points to the dire need to reform the contemporary 'military mind', mired as it is in the strategic literature of the past. One needs only to review the discussions on Australian strategy blogs such as *Grounded Curiosity*, *The Forge*, *The Central Blue*, and *The Cove* to observe how contemporary understandings of the military's utility to International Relations are firmly anchored to a small number of seminal works, many of which were published over 100 years ago. That much of the content of these blogs is generated by active military officers suggests two awkward truths. First, despite a consistent pattern of recent strategic failure (Vietnam and Afghanistan are perhaps the most obvious examples), military education programmes display a reluctance to broaden the aperture beyond texts that serve to reinforce a particular self-identification of military utility that is confined to high-end warfighting. Second, there exists a significant lack of academic interest in broadening the current strategic discussion, demonstrated by the near absence of any discernible cohort of scholars in Australia researching the military in disciplines such as sociology or philosophy. These two observations suggest that, rather than seeking new theoretical frameworks to inform military utility and operations in the face of today's complex, interconnected strategic environment, there is a broad consensus that extant strategic thinking is sufficient. Again, the lessons of Vietnam, Iraq, and Afghanistan cast doubt over such an assertion. In a fitting close to the section, Amy Hestermann-Crane drags the reader into the present with a concise history of Australia's engagement in one of the newer military domains: space. As with its similarly ethereal cousin, cyberspace, Hestermann-Crane argues that space presents 'new beginnings'. However, it remains to be seen whether military strategic thought will evolve to explore these opportunities or seek to bend them to its established tenets.

The final section of the book, much of it written by relatively junior members of the ADF, offered solace by demonstrating the existence of a healthy culture of debate within the ranks. Keirin Joyce, an Air Force officer who began his career in the Army, presents a cogent argument for greater focus on the utility of UAS to the ADF. Drawing on his unique experience working on UAS projects both in the Army and the Air Force, he is well-qualified to comment on how individual service cultures inform their view on the role and application of UAS technology. His conclusion – that the ADF needs to take a more integrated approach to implement such systems – reinforces the strategic direction outlined in documents such as *The Air Force Strategy* (2020) and the *Defence Strategic Update 2020*. Even more encouraging is the inclusion of *all* aspects of air power in the most recent iteration of The Air Power Manual (APM).[6] First published in 1990 – notably, one year after the decision was made to transfer ground-support rotary-wing assets from the RAAF to the Australian Army – the APM conspicuously omitted any reference to rotary wing assets as constituting 'air power'. In essence, this traditionally made the volume more an 'Air Force Manual' since it articulated how the RAAF's air assets contributed to joint strategic effect, rather than viewing air as a joint warfighting domain. The rise of space power as inherently joint, combined with a more inclusive definition of 'joint operations' as articulated in the latest edition of the APM, set a new horizon for how the RAAF self-identifies within the framework of ADF joint operations.[7] This point is reinforced by James-Andre Galam, who argues a strong case for greater focus on resilience for Australia's space capabilities, given their fundamental importance to the ADF's operational effectiveness. The call is picked up by Christopher Wooding in the final chapter of the book, in which he reviews emerging and disruptive technologies and their implications for the RAAF, concluding, fittingly, that the Air Force must remain at the forefront of integrating these technologies if it is to remain relevant and operationally effective into the 21st century. That these final two chapters were written by the two youngest contributors to the volume – indeed, the chapters trace their genesis to presentations made while both authors were still cadets undertaking initial officer training – bodes well for the future of air and space power in Australia. Such enthusiasm and critical enquiry will be fundamental to achieving national strategic effect with Australia's air and space power capabilities. It is hoped that this volume represents a renewed effort across the Australian air and space community to develop robust theoretical, strategic, and operational bodies of knowledge in support of this important work.

Notes

1 Thomas Kuhn, *The Structure of Scientific Revolutions* (Chicago: University of Chicago Press, 1970).

2 Chris Clark and Sanu Kainikara, eds., *Pathfinder Collection – Volume 4* (Canberra: Air Power Development Centre, 2009), 91–94, https://airpower.airforce.gov.au/sites/default/files/2021-03/PFV04-Pathfinder-Collection-Volume-4.pdf.
3 Colin Grey, *Airpower for Strategic Effect* (Maxwell Air Force Base, Alabama: Air University Press, 2012), 69.
4 Jarrod Pendlebury, " 'This Is a Man's Job': Challenging the Masculine 'Warrior Culture' at the U.S. Air Force Academy," *Armed Forces and Society* 46, no. 1 (2020): 163–84.
5 Royal Australian Air Force, *Air Force Strategy: Key Highlights* (Canberra: Commonwealth of Australia, 2020), 13.
6 Air and Space Power Centre, *The Air Power Manual*, 7th ed. (Canberra: Commonwealth of Australia, March 22, 2022), https://airpower.airforce.gov.au/publications/APM7thEd.
7 Grey, *Airpower for Strategic Effect*, 69.

References

Government sources

Air and Space Power Centre. *The Air Power Manual*. 7th ed. Canberra: Commonwealth of Australia, 22 March 2022. https://airpower.airforce.gov.au/publications/APM7thEd.
Royal Australian Air Force. *Air Force Strategy: Key Highlights*. Canberra: Commonwealth of Australia, 2020.

Published sources

Clark, Chris, and Sanu Kainikara, eds. *Pathfinder Collection – Volume 4*. Canberra: Air Power Development Centre, 2009. https://airpower.airforce.gov.au/sites/default/files/2021-03/PFV04-Pathfinder-Collection-Volume-4.pdf.
Grey, Colin. *Airpower for Strategic Effect*. Maxwell Air Force Base, AL: Air University Press, 2012.
Kuhn, Thomas. *The Structure of Scientific Revolutions*. Chicago: University of Chicago Press, 1970.
Pendlebury, Jarrod. " 'This is a Man's Job': Challenging the Masculine 'Warrior Culture' at the U.S. Air Force Academy." *Armed Forces and Society* 46, no. 1 (2020): 163–84.

Index